THE NEW NATURALIST
A SURVEY OF BRITISH NATURAL HISTORY

PESTICIDES AND POLLUTION

THE NEW NATURALIST

PESTICIDES AND POLLUTION

Kenneth Mellanby

COLLINS
14 ST JAMES'S PLACE, LONDON

First Edition 1967
Second (Revised) Edition 1970
Reprinted 1971

SBN 00 213177 3

© *Kenneth Mellanby, 1967*
Printed in Great Britain
Collins Clear-Type Press
London and Glasgow

CONTENTS

ILLUSTRATIONS

EDITORS' PREFACE

The *New Naturalist* series, now some way beyond its half century of volumes and its quarter century of years, has had much cause for gratitude to the senior officers of our kingdom's Nature Conservancy who have so valuably contributed to its books and its task.

It could be expected that a servant of a statutory body, when discussing problems of conservation and ecology of political, economic and social moment (as nearly all such problems are) might adopt a somewhat statutory tone of voice. None of our Natural Conservancy authors has yet done so; nor has Dr. Mellanby, who handles in this book what can be vulgarly described (if it is not mixing a biological metaphor) as the hottest potato in the nature business. The impact of modern industry's chemical products (themselves the products of vastly expensive and brilliant research) upon our environment—the air we breathe, the water we drink, the food we eat, and the animals and plants with which we try peacefully to coexist—is a subject so vast, so emotion-rousing, so socially provoking that almost anybody could be forgiven for approaching it with fearful frenzy on the one hand, or with careful dissemblance or even dissimulation on the other. Not so Mellanby: this is a calm book, and a deeply thought-out book, and patently a balanced book: the kind of book we expected from the leader of one of the finest teams of ecological analysts in the country. Of course we knew it would be so, when we persuaded the Director of the Monks Wood Experimenal Station to write it.

Kenneth Mellanby's approach is magnificently lucid, the more so because of his deft use of illuminative detail, alternating with wise generalisations that show the deepest understanding of the history of pollution and the eternal struggle of man against predators and pests. He has had to specialise in being general: be not only historian but geographer, physiographer, chemist and physicist as well as biologist to arrive at a sense of proportion and balance, and a true evaluation of the present tides and streams of wastes and poisons, their natural history, control, cause, cure and care.

This book has been written without fear or favour, and with the fairest analyses of mistakes and successes. It persuades us of the need for everybody's deeper understanding of the problems involved, and that our human stock, with the increase of its population and its civilised wants, has courted risks, certainly invited disasters and suffered a few—and yet may have succeeded in arriving at a point of common-sense confidence in a clean (or cleanish) planet in the predictable future. The planet is presently a pretty dirty one: but some clever ecologists and conservationists have voices that are now being heard, and may be heeded before it is too late.

All books on pesticides and pollution must be compared with the late Rachel Carson's classic *Silent Spring* of 1963, which started the general public of all the educated world wondering. Kenneth Mellanby's book is not of the same genre, and it would have been most inappropriate if it had been. Miss Carson's book was a chamber of horrors, and, as regards the insecticidal events of its time, as every responsible naturalist (including Dr. Mellanby) would agree, accurate. It did a power of good. This book, we predict, will do further good; for it does the next thing. It does not say *how awful!* for this has already been said, and in Rachel Carson's context justifiably. It says *how does the business really work, and what next?* and proceeds to spell it out, in a masterly style and depth that we are proud to be associated with.

THE EDITORS

AUTHOR'S PREFACE

In this book I have tried to deal with the subject of environmental pollution in Britain in an objective way. The public, and particularly those members of the public who are interested in the conservation of wild life, are very familiar with many types of pollution, but they cannot always judge their importance. Sometimes atomic radiation seems to be all-important, particularly when questions of military strategy are discussed. Our many fishless rivers are clear evidence of the serious effects of the pollution of fresh water, and every year our newspapers have pictures of dead seabirds covered with oil on our beaches. Great publicity has been given to agricultural chemicals. The remarkable impact of Rachel Carson's *Silent Spring* has suggested that insecticides are the greatest danger. I have deliberately avoided dealing with this book in the text, not because I underrate its contribution to the subject, but because I think that the time has come to try to look at all sides of the problem. Rachel Carson, when dealing with insecticides and herbicides, was careful to give us the facts as they applied to the United States, but she selected her facts, and gave us an advocate's case. At the time, this was a useful service to science, and equally selective rejoinders from the chemical industry have done little to reassure the public. Other and more objective books on the effects of pesticides are listed in the bibliography. One of my main tasks has been to try to fit pesticides into the general picture of pollution from all sources.

The writer of a book like this needs to call on many others for help. Over a good many years I have discussed these problems with scientists from many countries, and I have tried to digest their views and the contents of their publications. The question

13

of pesticides must have special mention. I have been fortunate in being able to discuss these problems with Dr. N. W. Moore, head of the Nature Conservancy's Toxic Chemicals and Wild Life Section at Monks Wood Experimental Station, and with the members of his team. Dr. Moore was the first scientist in Britain to organise research work on this subject, and he and his colleagues have made major contributions towards the understanding of their problem. I have received immense help from them at all stages and have taken up a great deal of their time in detailed discussion. I hasten to add, however, that they are in no way to blame for any faults in my presentation of the subject.

The editors of this series have given valuable help. Sir Julian Huxley originally suggested that I should write this book, and made useful proposals as to its contents. Sir Dudley Stamp also gave me much encouragement, and the most courteous application of the spur whenever I fell behind my schedule. Without this I would never have finished the book. He read the manuscript and I shall treasure the appreciative letter he wrote me about it not long before his untimely death.

As I have already said, I have tried to make this book an objective account of pollution. I fear that I shall be attacked from all sides. In discussions I have been accused of exaggerating the dangerous effects of industrial processes and of beneficial agricultural chemicals. I have also been told that I play down the dangers of these substances. So long as the attacks do indeed come from these different quarters I shall not feel that I have entirely failed in my objective.

CHAPTER ONE

INTRODUCTION

This book is essentially an account of the way in which man is unintentionally contaminating his environment. This is a world-wide problem, but for the most part I am restricting my scope to consider the situation in Britain. I do this for two reasons, first because this treatment seems appropriate to a volume in the New Naturalist series, and secondly because although the problem is more acute in these small, developed and densely populated islands than in many other parts of the world, it is possible that what we do can help and guide others, and serve as a warning to prevent damage in the countries at present undeveloped which might otherwise accompany their economic development.

I think it is quite logical to bring together in one book the effects on our environments and on our wild life of *pollutions* which arise from urban conditions and from industry, and the rather special case of *pesticides*, which have recently been shown to constitute such an important contribution to environmental pollution. In all cases we are dealing with effects which were not deliberately planned by those who produce them. Pollution, including radiation, has grown as a side effect of the increase in human population, and of increased material productivity. Pest control began in quite a different way, but it is the side effects of modern pesticides, substances which could only have been produced and developed in an industrial system, and which may be effective weapons when properly used, which are causing so much concern.

In this book I have attempted to review the present situation, show how it has arisen, and discuss how man may either continue to pollute the earth and to impoverish it by destroying and restricting its flora and fauna, or alternatively may control his own activities for the benefit of posterity. The situation is serious. Much irremediable damage has already been done. On the other hand, we have learned how to avoid the worst types of pollution. Public opinion is strongly in favour of controlling pollution, and we are not so dogmatically in favour of eradicating any forms of life we consider to be "pests" as perhaps we used to be. If mankind is prepared to make a determined effort, to support much more research to make that effort effective, our descendants may not be condemned to live on an impoverished planet devoid of so much of the varied life which has made it so interesting and so beautiful. There is one other point which should be noted. Some pollutions, particularly from insecticides, seem to be harmful to wild life but not, as yet, to man. Air pollution is more often seen to damage plants than to harm man. Perhaps we should look upon these as useful "early warning" systems, and use the opportunity to reduce the chances of future damage to our own species.

Man deliberately alters the face of the earth, to a degree to which no other species of animal has aspired. He builds cities, factories and roads which wipe out wild life over large areas. He does this quite intentionally, assuming that the gain far outweighs the loss, though his pollution of the surrounding countryside with domestic and industrial effluents is not intentional. However, important as are the effects of urban and industrial development, it is as a farmer that man has the most profound effects on the landscape, and on the plants and animals which live there. An excellent account of these changes is given in *Man and the Land* by Sir Dudley Stamp. Farmers destroy the natural vegetation, and substitute wide areas of alien plants grown in monoculture. They deliberately try to get rid of plants which compete with their crops – that is, weeds. They try to destroy animals which eat their crops and so reduce the amount harvested – that is, pests. As agriculture has developed, new methods of dealing with weeds and pests have been discovered and put into use. To-day we are all conscious of the fact that chemical controls have been developed, and that poisonous substances are widely used to achieve control. We know that these poisons may kill birds and other forms of wild life, and we fear that they may endanger human health. However, we tend to forget that these possible toxic side effects may be less catastrophic to wild life than are the ecological effects of "traditional" agriculture.

Man has existed for hundreds of thousands of years, but it is only within the last ten thousand that he has had any considerable ecological effect, and only within the last millenium that he has really begun to change the whole appearance of the earth's surface. Until about 8,000 B.C. man was a hunter and a gatherer of wild plants, with no domestic animals and no crops. Life, as Hobbes has said, was "nasty, brutish and short." Man had little effect on his environment. He ate fruit and berries, and helped to disperse the seeds as do birds and animals to-day. His waste-products were probably dispersed over the land and helped to fertilise it. When he died his remains decayed and the nutrients were returned to the system. The nastiness of his life was probably made worse by pollution, even at that stage of his development, and pests contributed to its shortness. Some animals are naturally clean. They excrete in recognised latrines some distance from their usual resting places, and they do not leave the decaying remains of food in inconvenient places. Other animals, and primitive man was probably one of them, are not so hygienic, and if he lived in a cave it probably stank disgustingly. Even when he buried his dead he sometimes did so beneath his dwelling, just where decay would be most objectionable. Under these conditions many diseases must have been rife.

Early man probably thought of the larger carnivores as the most serious pests. He competed with them for animal food when he became a hunter, and he himself formed a part of their prey. At first he did little positive to control these pests; his main object was to avoid them. Later, but long before he became a farmer, man in some parts of the world learned to trap and hunt even the most ferocious wild animals, and he probably accelerated the extinction of several species. Recent work on the so-called "Pleistocene overkill" goes so far as to suggest that man exterminated nearly half the larger mammals

P.P. B

in Africa some 50,000 years ago, and that in North Africa (where man arrived much later) he similarly killed off at least sixty per cent of the species of large mammals around 10,000 B.C. If these conclusions are finally substantiated, they will have a profound effect on ecological thinking. Early man probably paid little attention to the smaller insect pests, lice, bugs and fleas, from which he no doubt suffered. He did not realise that these were not only a nuisance, but were also the carriers of diseases which were far more deadly than all the lions and tigers and snakes which he so greatly feared. At this stage man was simply an animal, competing with other animals, and doing little to upset the uneasy balance of nature.

There were cases where hunters profoundly changed the landscape. It is likely that North American Indians deliberately burned the forest, and so encouraged grassland which could maintain larger herds of buffalo. This could be considered an early example of wild life conservation! It must also have had profound effects on all the other animals and plants in the region. Incidentally, at a much later date, when the prairies were cultivated to grow cereal crops, the buffaloes became "pests" and were almost completely exterminated.

Primitive man suffered from pests and pollutions, even if he was not always aware of this. When he became an agriculturalist, he recognised the fact. Settled agriculture, with the growing of some crops, has gone on in parts of the world for perhaps ten thousand years, but extensive farming has only existed for about five thousand years. Arable farming is essentially a process where certain plants are encouraged, and others, which would compete with them, are discouraged. The unwanted plants are pests or weeds.

It is difficult to find a satisfactory definition of a pest, other than to describe it as a plant or animal living where man does not want it to live. The same animal may sometimes be treated as economically valuable, at other times as a dangerous competitor. Thus mink, escaped from fur farms in England, where they are prized, are considered as dangerous pests in other parts of the country. The same plant may be a valued crop at one time and a weed at another; an obvious example is the potato, for a few tubers, accidentally left in the soil when this crop is harvested, are troublesome weeds in cereals grown in succeeding years.

A great part of the energy expended in arable farming goes in weed control. Ploughing, harrowing and cultivation are all means of reducing weeds and their growth, as well as of making conditions suitable for planting crops. It is therefore somewhat ironic that weeds are important largely because man has produced conditions in which they flourish. Most weeds were rare plants before man became a farmer, and some are now uncommon or even extinct except on farm land. The history of weeds and their development has been fully described by Sir Edward Salisbury in his masterly book *Weeds and Aliens*, to which any interested reader must refer.

At different times farmers have used different methods of weed control, and different species of plant have been economically important. When wheat and other corn crops were sown broadcast, hand weeding was the only practicable method. When drilling in rows was introduced in the eighteenth century, it became much easier to keep crops clean. At the same time improved methods of separating crop from weed seeds were devised, so that sowing did not itself greatly contaminate the ground. Thus with clean seed and properly planted crops, cheap labour and comparatively simple horse-drawn machines kept the fields clean. At the end of the nineteenth century there was no serious weed problem for the good farmer in most parts of Britain. As agricultural wages rose, mechanisation was introduced, and some farming processes were improved, but many crops became weedier and weedier, so that different rotations had to be developed, not always with success. In recent years the situation, for the farmer, has been saved by the introduction of selective weedkillers. These have revolutionised agriculture, but have produced their own problems, as will be seen in chapter 6.

Parasitic fungi cause a great deal of crop damage. Early man was aware of some of these diseases, and the danger of eating corn made poisonous, for instance, by the fungus causing ergot, particularly in rye. However, many fungus diseases were not recognised as such, and their damage was accepted as a normal risk of farming, until the latter part of the nineteenth century. As shown in chapter 7, many fungus diseases are now controlled with little risk to other forms of life.

Farmers, from neolithic times onwards, were aware of mammalian pests, ranging from deer which damaged their crops to wolves which preyed on their herds. Rats and other rodents were known to consume much of the stored grain, and many ingenious methods of excluding them were devised. Early pest control was in effect hunting; the results were sometimes successful, as in the case of large and slow-breeding animals, and quite ineffective against small mammals which bred rapidly. Brown bears were exterminated in England in Roman times, and in Scotland before the Norman invasion. Wolves continued much longer. They were quite common, particularly in Wales, into the medieval period, and the last survivor is believed to have been killed in 1740; a few lingered on in Ireland for another thirty years.

Pest control in the English countryside has usually been related to game preservation and sport as well as to agriculture. The wolf was clearly too large and voracious an animal to be tolerated, and so it was eliminated. No one to-day seriously suggests its reintroduction, though conservationists (if not the local farmers) are concerned about its future in Spain. The fox, however, is still quite abundant, although it undoubtedly kills poultry. It would be difficult, though not impossible, to exterminate all the hill foxes in the wilder parts of Britain, but a determined effort could get rid of this animal in areas of intensive farming within a couple of years. Foxes may not be deliberately preserved, but they have been tolerated for many years because of fox-hunting, and those naturalists who are opposed to blood sports should realise their debt in this connection. Recent research has shown that foxes live as much on carrion and on small animals and insects as upon poultry and game, so they are likely to continue to survive unless accidentally wiped out by chemicals (see p. 140). Recently foxes have been reported in increasing numbers in suburbia, raiding dustbins and feeding on garbage. These habits will not endear them to the more sentimental members of the public, who may equate garbage-eating foxes with rats. More of the animals are likely to survive, and they may eventually become pests in a new role if they become too common.

Gamekeepers were for a long time the main enemies of carnivorous animals and birds, which they spoke of as "vermin" on the assumption that they lived mainly on game. Most keepers until recently had their "gallows" on which the rotting corpses of stoats, weasels, hawks and owls were hung, presumably *pour encourager les autres*. Systematic shooting of slow-breeding predatory birds effectively controlled their numbers. In the nineteenth century kites, formerly distributed throughout the country, were eliminated except from a few mountainous areas. A careful investigation by Dr. N. W. Moore shows how the buzzard has fared in the last 150 years. At the beginning of the nineteenth century buzzards were quite common breeding birds over most of Britain. By 1865 they had been exterminated in most of East Anglia and the Midlands, and by 1900 they were only to be found in Cornwall, Wales, the Lake District and Western Scotland. By 1954 the situation had somewhat improved, and the birds had recolonised many of the areas occupied in 1865. This spread was clearly due to the decrease in game preservation during and after the 1914-18 war. If the data for buzzards and the numbers of gamekeepers are mapped side by side, this shows a good negative correlation.

To-day many gamekeepers are more enlightened. Although hawks and other carnivorous birds do eat some game birds, they prey much more on small mammals (mice, voles, etc.) which compete with game for food. Most predators are now themselves legally "preserved," though they can be shot if caught in the act of taking poultry or game. We may see an improvement in numbers, similar to that manifested by the buzzard between 1900 and 1954, for many other species, if the new danger from agricultural chemicals can be overcome.

Smaller, non-carnivorous mammals and birds do a lot of damage, the amount of which is not always recognised. Rabbits

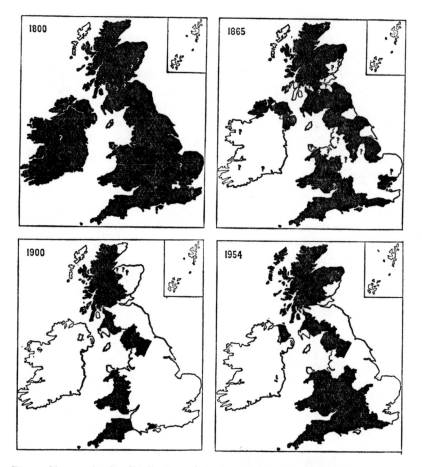

Fig. 1 Changes in the distribution of the buzzard in the British Isles.
(from Dr. N. W. Moore with acknowledgement to *British Birds*).
KEY: *Black:* Breeding proved, or good circumstantial evidence of breeding.
? *on black:* Circumstantial evidence suggests that breeding probably took place.
? *on white:* Inadequate evidence of breeding.
White: No evidence of breeding.

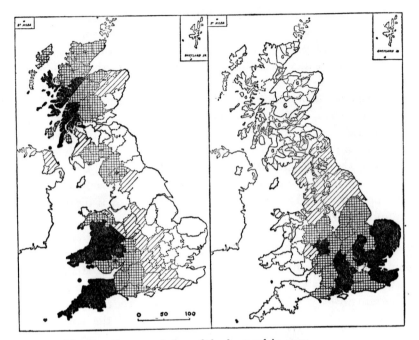

Fig. 2 *a.* The breeding population of the buzzard in 1954.
KEY: *Black:* 1 or more pairs per 10 square miles.
Cross-hatch: More than 1 pair per 100 square miles, but less than 1 pair per 10 square miles.
Diagonal hatch: Less than 1 pair per 100 square miles.
White: No breeding buzzards.
+ means that breeding density may belong to the category higher than that indicated.
— means that the breeding density may belong to the category lower than that indicated.

b. Game preservation in 1955.
KEY: *Black:* 3 to 6 gamekeepers per 100 square miles.
Cross-hatch: 1 to 2 gamekeepers per 100 square miles.
Diagonal hatch: Less than 1 gamekeeper per 100 square miles but more than 1 per 200 square miles.
White: Less than 1 gamekeeper per 200 square miles.
G. Principal grouse-preserving areas. On these, and also on some very large estates, the numbers of keepers may be higher than shown on this map.

were probably introduced into England by the Normans in the eleventh or twelfth century and were comparatively uncommon, prized as game animals until the nineteenth century. Then, for some reason which we do not yet understand, they suddenly increased in numbers so that a density of twenty animals to an acre was not uncommon, and as many as 100,000,000 carcasses were sold in a year, without noticeably depleting the numbers still at large. It was not until rabbits were wiped out in many areas by myxomatosis in 1954 that the extent of the damage they had done was recognised. Pre-myxomatosis control, by shooting and trapping, was rarely effective, and merely served to "crop" the population. The importance of myxomatosis is discussed in chapter 10.

Mice and rats invaded man's home as soon as there was a home to invade. They also lived in his grain stores and farm buildings. Early man tried to make his granaries rodent-proof, sometimes with remarkable success. Control by trapping and poisoning was generally inefficient, and the rodent population in contact with man was roughly a measure of the amount of food he made available. A mouse-proof larder is more effective than an apparently efficient trap. The domestic cat was probably the most useful method of local control, though cats, like other "pesticides," have had their side effects. Those that have escaped and become feral have important effects on other wild life. Modern methods of rodent control are much improved, but these animals still do much economic damage in our cities and on our farms to-day. It is perhaps surprising that the black rat (*Rattus rattus*), which is found mainly in towns, where it is particularly at home in hot-water ducts in tall buildings, was our "original rat," though some think even it only arrived about A.D. 1200, and the brown rat (*Rattus norvegicus*), which is the species commonly found in the country, only arrived in Britain in the eighteenth century. The black rat was apparently driven from the rural haunts by the brown invader.

Pigeons and sparrows are serious agricultural pests, against which no really satisfactory control measures have so far been devised. Pigeons become more numerous each year. This increase is probably due to the increased amount of winter food, particularly clover leys on farms, that is available. The recent fall in the number of hawks may also have had some effect. Organised shoots give some sport to the participants, but have negligible effects on the pigeon population. Work on poisoning or narcotising pigeons is progressing, but the danger to other and more desirable species of birds is difficult to prevent. Sparrows probably increase because suburban householders feed them in winter. This enables them to survive in cold weather, and the increased population does more harm on the nearby farms to which it migrates in summer and autumn. Again shooting and trapping has little effect.

Farmers and others were then soon aware that wolves and other large mammals and birds might be pests, competing with them in various ways, even if they often overestimated the damage done by the carnivores and sometimes underestimated the amount of food taken by rabbits and other herbivores. They took active, if sometimes misdirected, steps to control these animals. On the other hand, they almost always underestimated the harm done to their crops and to their health by insect pests, and it is only in recent years that serious attempts have been made at control in this direction.

In Britain we do not have plagues of locusts, which in many countries can consume the whole of a crop, but it is estimated that to-day some £300,000,000's worth of food is lost each year because of pests of crops and insect damage to farm stock. This sum is almost exactly the same as is spent annually on the support of agricultural prices ("farm subsidies"). Some insects, like the caterpillars of Cabbage White butterflies, eat crop plants, reduce the yield and make many plants unsaleable. Other insects do little damage themselves but carry organisms which cause diseases. Thus aphids carry the virus causing virus yellows in sugar beet; this can seriously reduce the value of the crop. For many years farmers and gardeners accepted insect damage as something they could not prevent. They learned by experience that it could sometimes be avoided or reduced to a minimum by timing their operations carefully. Thus if broad beans in the garden are sown early, seed is set before aphids ("black fly") are numerous and a good crop is obtained, but in two years out of three a late sown crop will be smothered by insects and prove a failure. Cultural devices such as this are valuable and important, but will seldom allow a late crop of broad beans to be produced. Field beans, which flower and set seed over a long period during the summer, cannot be successfully grown in a way to avoid attack in a bad aphid year. Resistant strains of crop plants have been recognised and used for many years, and in future are likely to prove very important, where resistance is linked with high yield and quality. Farmers generally prefer to be able to grow the most profitable crop at the most convenient time and wish to attack pests by any possible means. Before 1939 most effective insecticides (except general poisons very dangerous to man) could only be produced in relatively small quantities and at a price which made their use on many crops uneconomic. The farmers therefore have welcomed the synthetic insecticides which can be produced in unlimited quantities and which, at first at any rate, seemed the perfect answer. The dangers from their use are described in later chapters.

Insect pests of crops were soon recognised as such, even if little was done about the problem until very recently. The importance of insects as vectors of disease has only been understood for about seventy years, though man and his habitations have provided niches for troublesome parasitic species in the same way that wild animals have supported their own parasites. Man has sometimes controlled insects of medical importance effectively without understanding the problem. Malaria was formerly widespread in Britain, but it was almost eliminated long before man knew the parasitic organism concerned or that it was carried only by the *Anopheles* mosquito. This was because man avoided marshy areas, thinking that malaria was caught from the "bad air," and so he kept away from the breeding places of the mosquitoes. He also drained the swamps, usually to produce better agricultural land, but in so doing he got rid of the insects which carried the disease. Incidentally naturalists are now very concerned at the continued draining of marshes and swamps, which are now the last refuges of many species of wild life. This is just one of the ways in which non-chemical pest control can have effects which have end results which may be as disastrous to wild life as the most indiscriminate use of chemicals.

In the Middle Ages most people of all ranks of life harboured body lice on their persons, and as recently as 1940 the majority of the girls in our industrial cities had lousy heads. Personal cleanliness could eliminate these pests, except where conditions were grossly overcrowded, but in war and after any disaster infestation, and the risk from louse-borne typhus fever, grew. Persistent insecticides now control these insects, and properly applied to man and his clothing they present little danger to any other organisms.

Pest insects which attack man, but depend on his having a permanent home, include fleas and bedbugs. Fleas are not only a nuisance, but also carry plague, a disease which died out in Britain many years before effective insecticides were discovered to control the vectors. Improved hygienic conditions rather than chemicals have made fleas uncommon insects.

Human bedbugs are very similar to the species which attack bats and swallows, and primitive man may have become infested first when he also lived in caves. Bugs were common from the earliest times in the warmer parts of the world. However, Britain had no bedbugs before the sixteenth century, possibly because the houses were too cold. Bugs certainly existed in Italy in classical times. When bugs arrived in Britain, they soon spread through overcrowded slums, and before the 1939 war most houses in our cities, except detached surburban villas, harboured at least a few. However, they were only common in unhygienic and overcrowded dwellings, and improved conditions soon reduced their numbers. Modern insecticides, particularly those which put a persistent film in the cracks which the bugs haunt, control the insects effectively without seriously contaminating the environment.

When his numbers were few, pollution was not a serious problem to man. Many pests, both plant and animal, have become common only because man has produced suitable conditions. In some cases pests have been controlled with little harm to the environment, in others pest control has become a new and potent form of pollution. The great difficulty is to assess accurately just how much pollution affects the environment and the plants and animals it contains. We are seldom able to give simple answers. Sometimes an animal has obviously been killed, perhaps from the effluent from a factory, perhaps by accidental contamination with a pesticide. Generally, however, we have to depend on circumstantial evidence of damage, and this is the reason for the controversy which so often surrounds our subject.

Toxicology is difficult and complicated. The results of analyses of animals' bodies, where traces of poisonous substances are found, are not easily interpreted. In the case of well-known poisons like arsenic, strychnine or cyanide the situation may be less mysterious if large amounts are found. We know, for instance, that if a man eats five grams of lead arsenate he is likely to be fatally poisoned. If the pathologist finds ten grams of lead arsenate in the stomach of a corpse, he will be almost certain that this poison caused death. If he finds only a few milligrams, he will be almost certain that death was due to some other cause. The finding of intermediate amounts makes diagnosis difficult. Consideration must be given to the site where the poison is found in the body, and to losses due to vomiting or excretion. The situation is even more complicated where poisons are broken down in the body, either as part of the process of damaging the victim, or due to post mortem changes. If we do not know accurately how toxic a chemical is to a particular animal, and if we are not fully familiar with these chemical changes, we cannot usually say for certain whether a small residue of poison in a live or dead specimen has any significance.

It is generally fairly easy to establish the acute toxicity of a substance, that is the amount which, in a single dose, is lethal. Experiments with rats, chicks or fish are commonly made. Groups of animals are given different doses, and the least amount of poison which kills is found. Usually different individuals of a species show a somewhat varied susceptibility, and instead of determining the amount which kills them all, the so-called LD_{50}, that is the amount which kills half of a batch, is determined. In most instances few animals die from a single dose of half the LD_{50}, and twice the LD_{50} is likely to kill almost every individual. However, this is not always the case. Sometimes a population contains a few individuals which can survive relatively large doses of certain poisons; under certain circumstances these may be selected out and may breed a strain which is more resistant than the normal to a toxic substance. Resistance or susceptibility to poisons is not necessarily correlated with unusual or subnormal "vigour," and this type of chemical selection may leave a species less well adapted to normal environmental conditions.

Although acute toxicity is not difficult to determine in the laboratory, it can only be done with a limited number of species, and values for others (including man) can usually only be inferred. Also the effects of a specific poison may differ even with the same batch of the same species depending on how it is administered, e.g. neat, in suspension, in oily solution, on an empty stomach, through the skin, by inhalation and so forth. These difficulties have usually meant that, at least where man is exposed, a fairly large "safety factor" has been applied. Thus if work with rats suggests that the LD_{50} for substance "X" is 50 milligrams per kilogram, it could be assumed that half of a group of 50 kilogram men would probably die if they ate one gram each of "X." It would generally be found that a single dose of one hundredth of this amount, i.e. of 10 milligrams, would be unlikely to be harmful. In many cases this assumption is quite justified but contamination of food to this extent would not normally be tolerated.

While there are sometimes difficulties in establishing the effects of single, large doses of poisonous substances, the study of the effects of repeated small doses, each of which would probably be harmless, spread over long periods, presents even more serious problems. Poisons which are unstable are unlikely to be very dangerous under these circumstances. Those which are stable, particularly if they are stored in the body, may present great risks even if they are not acutely poisonous in single doses. All these factors are borne in mind when, for instance, new insecticides are tested. Their action on a number of insects, particularly pests, is determined. Then long-term experiments, lasting over severa, years and a number of generations, are then made with rats, chickens and other animals. It is obviously impossible to include more than a few species in such trials, so it is not surprising that sometimes a desirable species of bird, or mammal, is found (too late) to be unexpectedly susceptible. The effects of chronic exposure to low-level industrial and urban pollution is even harder to study. Some, impressed by the complexity of the situation, fear that the ecological effects of pesticides may bear little relation to their gross toxicity.

Everyone wishes to abolish the damage which may be caused to man and to wild life by pollution from every source. As, however, we are not always agreed as to when damage is being caused, or how exactly some obvious damage arose, an easy solution will not be found. Man has always polluted his environment; he has always suffered from pests, but because of the "population explosion," these problems have become more serious in recent years. The need for more research in these subjects is obvious, if irreparable damage to wild life, and to man, is to be avoided. Equally important, we must make sure that the results of such research are quickly and efficiently applied.

CHAPTER TWO

AIR POLLUTION

Perhaps the most obvious way in which man has contaminated his environment is by polluting the air with smoke and with the waste products from industry. Everyone has seen the pall of smoke hanging over a city. He knows that many plants and animals are not found in the middle of a city. It is, however, difficult to find exactly how this pollution has affected wild life, notwithstanding much intensive study of the subject. Although some lichens and other plants seem to be particularly susceptible to the effects of atmospheric pollution, and their distribution may be correlated with it, nevertheless the position is far from simple. This is perhaps not surprising, as we seldom have a constant amount of any noxious substance in the air at any place over any long period of time. The smoke emitted from a domestic fire or from a factory is in bursts followed by periods of comparative inactivity; in some towns factories are only allowed to give out black smoke for five minutes in an hour. The weather has a profound effect; calm clear periods, particularly when temperature conditions prevent upward circulation, allow the pollution to concentrate, while strong winds ventilate the area though they carry the substances in detectable amounts to distant parts of the country.

As soon as man discovered fire, he made smoke and so polluted the atmosphere. The effects were local and slight until about the thirteenth century, when coal fires in cities were found to produce winter fog and punitive laws were introduced, apparently with little permanent effect. As cities were small, and little coal was burned, probably no great damage was done except perhaps to men themselves living in unventilated houses. When cities grew, smoke became, and still is, a major problem.

Industrial development in the nineteenth century was accompanied by new types of pollution. Hydrochloric acid gas from alkali works caused a public outcry, with resulting legislation. Attempts have since been made to restrict all the emissions from factories to a "safe" level. This happened none too soon. Much of the gross pollution accompanying the dereliction in areas like the lower Swansea Valley was airborne from factories in the area.

The results of atmospheric pollution differ in an interesting way from those of insecticides which are discussed in later chapters. Man himself has been the major victim of polluted air; insecticides have had serious effects on wild life, but man has seldom been injured by the direct effect of these substances. The ecological significance of this difference is discussed in later chapters.

Every urban housewife is only too well aware of the reality of atmospheric pollution. Curtains and furnishings remain clean for months or years in the country; in the towns they are grimy in a matter of days. Students of pathology who have only seen inside the corpses of city-dwellers are amazed, and think they have found some new disease, when they see for the first time the healthy red lungs of a farm worker who has never lived in or near a town. Walkers on the moors of the Peak District know that their clothes will be blackened if they sit on the heather, and most flocks of sheep there, except immediately after shearing, seem to consist only of black sheep. The Peak District sheep on moors surrounded by industrial towns contrast with the much whiter animals found in the remoter highlands of Scotland, and this colour difference has been suggested as a rough and ready means of estimating pollution.

Air pollution in Britain to-day is mainly due to burning coal and oil. Local effects from many chemical processes, and petrol and from diesel engines also make their contribution. Perhaps the most serious chemical problem is due to fluorine, mainly from brick works, and this is specially mentioned below. Legislation and regulations have reduced the amount of many pollutions to such an extent that wild life is usually not seriously

1 Industrial dereliction in Britain. *Above*, the Lower Swansea Valley, where metal smelting in the eighteenth and nineteenth centuries has contaminated the soil with excessive amounts of copper and zinc. *Below*, a colliery tip at Mary Port in Cumberland. In some areas trees are beginning to mask such desolation, but the best solution is to put the waste material back underground

2 Air pollution by smoke and fog. *Above*, a winter morning in London. *Below*, industrial Sheffield

harmed, except in particular danger areas, but the amounts of dust, smoke and sulphur dioxide produced from fuel are so enormous and so unaesthetic that they cannot be ignored.

Britain consumes annually about 200,000,000 tons of coal and 25,000,000 tons of fuel oil. The output of noxious products is estimated at 1,000,000 tons of dust, 2,000,000 tons of smoke and over 5,000,000 tons of sulphur dioxide. Coal produces relatively more smoke and dust, and oil more sulphur dioxide. This pollution is obviously very unevenly spread over the country. The Ministry of Technology, formerly the Department of Scientific and Industrial Research, compiles reports from some 2,111 recording instruments spread all over Britain. These show that in heavily industrialised areas over 1,000 tons of grit and dust must fall on each square mile in a year; this corresponds to about two pounds on each square yard. In cities generally the figure is in the region of a quarter of a pound, and in rural districts it may be less than a tenth of an ounce. Sulphur dioxide, being a gas, is dispersed more readily, and the rural concentration is probably about a tenth of the urban or industrial figure, though under unfavourable conditions much higher values may be obtained adjacent to some factories. The housewife knows that polished silver or copper tarnishes more quickly in the town than in the country; this is correlated with the SO_2 in the air.

The effects of industrial pollution on man have been studied intensively, but with somewhat confusing results. It is believed that the four-day "smog" in December, 1952, killed some 4,000 Londoners. Exactly how smog, which is looked on as a brand of fog containing more contaminants and smaller and more penetrating particles, kills is not understood. It may act as a general irritant which acts as the "last straw" in the weak and those with respiratory trouble. It has been suggested that the excess of free sulphuric acid is the lethal factor, but total amounts are small (only 0·05 parts per million as a maximum) and this view is not generally accepted. There is no doubt that smog is a killer, and it kills other animals than man if they are exposed (many cattle died at the 1957 Smithfield Show), but fortunately it does not often spread outside our largest cities. Mist, which consists of relatively clean water particles, is of course widespread. Fog, which is essentially mist containing amounts of smoke, penetrates some distance from industrial areas, but seems to have comparatively little acute effect on man or animals.

Acute effects on man and animals of smog, and possibly of fog, can be shown to occur even if they cannot be fully explained. Chronic effects of the usual urban levels of pollution no doubt occur, but are not so easily demonstrated. Lung cancer is higher in cities than in the country, but we do not know the precise cause. Respiratory diseases are similarly commonest in industrial areas. Although we ourselves filter the air we breathe and reject much of the dirt, city dwellers' lungs are impregnated with dirt particles, and it is difficult to feel sure that this is not harmful. For these reasons considerable efforts are being made to reduce atmospheric pollution. "Smoke-free" zones have been scheduled in most cities, and some progress is being slowly achieved to reduce the smoke and dust. Fogs and smogs are less serious than they were, though the amount of sulphur dioxide in the air is less easily controlled and tends to increase even in smoke-free zones.

Farmers near to cities suffer from the effects of smoke and grime. It has been estimated that pollution, by damaging pastures in particular, costs the East Lancashire farmers over two and a half million pounds a year. Horticulturalists find that smoke reduces light intensity indoors and out, and obscures the glass of greenhouses, covering them with deposits which are difficult and costly to remove.

Smoke, by reducing light intensity, will obviously retard plant growth, and may encourage some species at the expense of others, though there seems remarkably little evidence of this happening except in industrial areas. Many city gardens do indeed suffer from the lack of light, but this is not due to pollution so much as to shading from buildings, and, more particularly, from trees. The luxuriant growth on bomb sites was a revelation to many. Here shading from buildings and trees was reduced to a minimum. Often one finds that spring flowers do quite well, before the trees are in leaf. In the confined space of a small city garden we may prefer trees to flowers, but we can seldom have both.

The effects of heavy deposits on leaves may be even more important. Evergreen species in heavily polluted areas have been shown to have a rate of transpiration of only about one tenth of normal and the leaves last a much shorter time than they do in pure air. Thus in some conifers the leaves normally live for up to eight years, and contribute to photosynthesis and growth for the whole of their lives. With moderate pollution the leaves may die and fall off in three or four years; heavy pollution may cause annual leaf fall and such trees hardly grow perceptibly and usually die. Some workers have suggested that particles of grime act by bunging up the stomata, but usually it seems that these are left patent and the effects are due to reduced transpiration, and, in some cases, to poisoning from sulphur or other substances. Deciduous trees which lose their leaves each year are often less susceptible to damage from pollution, as the leaves can complete their normal work before they are put, partly or entirely, out of action. Those responsible for planting in public parks in cities and industrial areas are well aware that spruce and firs are less likely to succeed than larch or oak. The exact way in which pollution harms trees is not fully understood.

I have already mentioned that atmospheric pollution by sulphur dioxide is becoming worse rather than better. The air in cities commonly contains 0·1 parts per million, that in rural areas 0·01 parts, but sometimes concentrations as high as 1 part per million may occur locally, under particular weather conditions, at distances from the source. Experiments have shown that most flowering plants show no damage to 0·1 parts per million even with long exposures, but higher concentrations usually cause damage such as leaf blotching and loss of yield. Some of the crop reductions on farms near towns are probably due to this cause, but it seems unlikely that there is much damage to wild life in rural areas. Sometimes this type of pollution may be economically advantageous; the absence from industrial areas of the fungus causing rose mildew is almost certainly due to sulphur in the atmosphere. This suggests that other species of fungi, which are in general much more susceptible to sulphur poisoning than are flowering plants, may be similarly affected. This could be of considerable ecological importance, but there seems little information on the subject. However, as rose mildew soon manifests itself in the suburbs, it would seem likely that other fungi, and susceptible plants of other groups, suffer little damage outside very polluted areas. Nevertheless it would be wrong to be dogmatic about this. Small quantities of sulphur or of other gaseous and solid pollutants which are dispersing through our environment may be more harmful than is generally realised.

It should be noted that although trees may suffer from the effects of pollution, at the same time they do something to alleviate this condition. It has been shown that trees growing in industrial areas may do this in several ways. They filter the air, so the deposits on their leaves are removed from general circulation. They cause turbulence and deviation of the air flow, which may help to ventilate (with less contaminated air) an area of otherwise stagnant pollution. They also remove carbon dioxide and liberate oxygen, an important function on a global scale, but, as mentioned below (p. 43) even seriously polluted air is almost never deficient in oxygen and dangerous concentrations of carbon dioxide are uncommon. Incidentally, in a highly polluted area where trees are likely to improve conditions, it may be best to plant broad-leaved deciduous species, which are harmed less than evergreen conifers, even though they have less effect in winter when the branches are bare. Conifers will be more efficient, and in some circumstances may be used and considered as "expendable."

Motor vehicles are responsible for widespread pollution in town and country. The exhaust gases contain a high concentration of carbon monoxide, which is very poisonous to mammals and birds. This gas may reach dangerous levels, particularly to car drivers in traffic blocks in towns, but it is probably dispersed too rapidly in the country to have an appreciable effect. Some three thousand tons of lead are emitted with the exhaust gases of cars in Britain each year. This has been found to accumulate in the vegetation and soil along roadside verges, and although serious damage has been seldom reported up to now, a dangerous

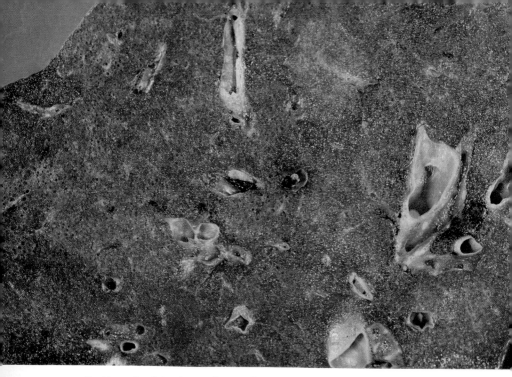

3 *Above*, healthy lung tissue of an 80 year old woman from Surrey, a non-smoker. *Below*, lung tissue of a 60 year old man, a smoker from Central London. The blackening is caused more by the dirty air than by smoking

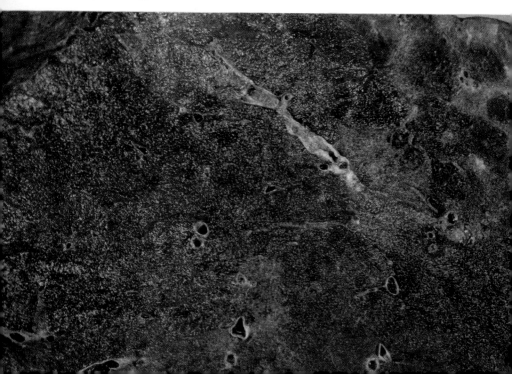

concentration could build up locally over a period of years. Lead could possibly enter food chains and have damaging effects far from the source of pollution.

Carbon dioxide is another common constituent of the exhaust from fires, factories and vehicles. It has seldom been found in the high concentrations which are harmful to life, and its presence may even promote plant growth in the way it has been shown to do when CO_2-enriched air is pumped into a glasshouse. Thus if the CO_2, which is normally only some 0·03-0·04 per cent of the total air, is increased to 0·15 per cent, the rate of photosynthesis in a glasshouse may be more than doubled, and crop yields can be substantially increased. The effects of CO_2 from industrial pollution on outdoor crops and on natural vegetation have not yet been thoroughly investigated. It is possible that quite small differences in CO_2 may affect the whole pattern of vegetation by stimulating one species of plant more than another. More work on this problem is clearly required.

Recently it has been suggested that CO_2 may eventually have a drastic effect on world climate. Coal and other "fossil fuels" are being burned at such a rate that the CO_2 content of the whole atmosphere may be raised as much as 25 per cent by the year A.D. 2000, and the level will probably continue to rise. The effects of this are not fully understood but some scientists think the temperature and other properties of the stratosphere may be affected. This could alter the world's radiation balance, possibly melting the polar ice cap. So far little or nothing has been done to reduce the output of CO_2, though some research on ameliorating its possible effects has been suggested. So far most scientists have thought that CO_2 pollution was of little importance; it now seems possible that it may cause greater changes to the world than any other man-made factor in our environment. On the other hand, this may be a completely false alarm.

Ozone, the form of oxygen with three atoms in the molecule (O_3) instead of the normal two (O_2), occurs naturally in tiny quantities, and pollution, particularly from motor vehicles, may increase the amount. As little as one part of ozone in 10,000,000 parts of air has been found, in the U.S.A., to harm many plants and trees, and such ozone poisoning is said to be important in both California and Connecticut, in which state an annual loss of $1,000,000 to vegetable crops is reported. So far, I know of no cases of ozone damage to vegetation in Britain, but with the increasing number of motor vehicles it seems likely to occur either now or in the near future.

Air pollution also affects the soil. Near cities the soil is often considered to be "sour," because of the sulphur dioxide and other acid-forming substances washed in by the rain. This effect probably does not extend very widely, but many of the chemicals found in rain-water may come from industrial pollution. In the moorland areas of the Pennines we know that the rain brings in substantial quantities of minerals, which contribute to the fertility of the soil. Much of this comes from the ocean, but some from pollution, which here may be having an advantageous effect. The quantities of nutrients are significant, but probably not sufficient to have detrimental effects such as those produced by similar nutrients in much larger amounts in purified sewage, which upset the balance in many rivers (see p. 52).

Botanists have studied the effects of pollution on a wide range of plants, mostly with inconclusive results. They have attempted to find "indicator species" which may be used to measure pollution. Such a species would only grow where pollution was below a certain level. The most successful work has been with lichens. Several species of lichen are absent entirely from the industrial areas of high pollution, and reappear on the outskirts. This problem has been studied in Northern Ireland, near Belfast, and around Newcastle upon Tyne. Fig. 3 shows how the lichen cover of tree trunks increases from the city centre of Belfast to its outskirts. It has been reported that the habit of growth of individual species was affected, so that some seemed barely able to exist where others grew normally. The subject is, however, not an easy one. It is necessary to be competent to recognise individual lichen species accurately, and to distinguish these in their sterile sorediate forms which often occur under unfavourable conditions. I must confess that I personally have been disappointed by the potentialities of this group. After reading in the

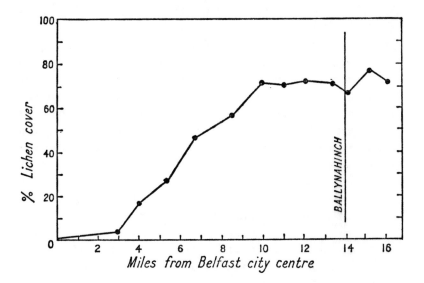

Fig. 3. Increase of lichen cover outside the city of Belfast. (After A. F. Fenton.)

literature that "a salient feature of lichen's ecology is that these plants are very scarce in the neighbourhood of towns" I visited the Lower Swansea Valley (Plate 1), perhaps the most polluted area in Britain. My first impression of the soil was an almost pure culture of lichen, and a wooden railway bridge was equally encrusted. These were of course resistant species easily recognised by a specialist, but showing that the method used in Belfast and illustrated in Fig. 3 cannot be generally used except by experts. In time botanists may find other plants which are better indicators; in fact there has been some progress in this field, but the confusion to-day relating to the effects of pollution suggests that unless it is very severe it may not be a factor of major importance in the ecology of most regions.

One particular element – fluorine – requires special attention. Fluorine occurs in minute quantities in all plants and animals, and it is one of the essential elements of protoplasm. If the natural level falls below a minimum, and this occurs in nature, harmful effects may be seen. One (but only one) of the reasons for the poor teeth found in many parts of Britain and North America is that the natural water may have a very low fluorine content, less than one tenth part per million, and combined with a "sophisticated" diet this may cause fluorine deficiency. A tiny additional amount, up to 1 part per million, may then be added to the water, and this has been found to improve tooth structure in children and reduce dental decay. This is one instance of a general principle, that a substance essential in small amounts may be toxic when the proper level is exceeded. The toxicity of fluorine in larger doses has made some people oppose the addition of this element to water, though there is no evidence that drinking water with 1 part per million ever does harm.

Fluorine occurs particularly in the smoke from brickworks, which are often surrounded by agricultural land. Other industries, including iron and aluminium production, are also important in this connection. Unlike active organic poisons, which may break down quickly to harmless substances, once fluorine has contaminated an area it remains a danger until it is physically removed. Fluorine only damages plants at relatively high concentrations, though it is at least ten times as toxic as sulphur dioxide. However, phytotoxic concentrations are rare, even near to industrial sites. The main danger from fluorine is that after deposition it is concentrated by growing plants. For example, grass has been found with as much as 2,000 parts per million. If this grass is eaten by stock, or by wild animals, they will certainly be seriously affected and will probably be killed. Lower concentrations have less drastic effects. The first symptoms of fluorosis are dental; the teeth are rough and mottled. Bigger doses cause bone abnormalities, lameness and general loss of condition. I know of no reports of fluorosis affecting wild life, but small mammals in affected areas are certain to suffer. Its stability, and the way it is concentrated by many food plants, makes fluorine a potential danger anywhere near a source, and abnormal weather conditions and air currents could affect vegetation, and thus animals, over a wider area. Fluorine seems a rather special, and dangerous, case of a poisonous substance entering the atmosphere, but it should make us more careful about accepting pollutions which may contain other, as yet undetected, dangers.

Air pollutions can thus have acute effects, when intense in industrial regions. They can have chronic effects, which may extend further from the source. In these cases emissions are acting as poisons, and the effects depend on the susceptibility of different plants and animals. In general, wild life, being remote from industry, would seem to be little harmed. However, there is one other way in which air pollution affects wild life, indirectly, by altering the physical environment.

We have noted that as much as two pounds of dirt may be deposited in a year on a square yard of ground near a factory. On the outskirts of our towns, the amount is perhaps an ounce. An ounce is quite a large quantity, more than the weight of pigment necessary to turn a blank paper into a valuable painting. Sheep within a considerable distance of industrial towns are black, and so are tree trunks and most animate and inanimate surfaces after a few months' exposure. We find that a number of different species of moths, which are normally pale coloured in unpolluted districts, are usually represented by *melanic* forms which are black or at least much darker than the "normal." This phenomenon of industrial melanism has been fully reviewed by Dr. E. B. Ford in his book *Ecological Genetics*, so there is no need to go into details here. It has been established that various moths, and the Peppered Moth (*Biston betularia*) has been most fully studied, have evolved melanic races which are adapted to their new surroundings. In clean areas, where tree trunks are covered with pale lichens, the typical form of the Peppered Moth is difficult to see. The melanic form is very prominent. This difference is not only apparent to man, but to birds which prey on the insects, and readily take them when resting on trees. In industrial areas, where the trunks are blackened and lichens are comparatively scarce, the melanic form is inconspicuous and is preyed upon least. This phenomenon has demonstrated that evolutionary changes may be more rapid than had previously been imagined. Not all evolutionary changes have such obvious morphological differences as we find in the Peppered Moth, and differences in physiology or behaviour may be selected and perpetuated by pollution, with important effects on wild populations which may spread outside the area in which they first become apparent. Thus many types of organism may be changing to-day, as a result of industrial pollution, with far-reaching effects which we do not yet suspect.

Man-made air pollution occurs where man is most numerous, so we are the species most affected. For this reason we take many steps in the attempt to safeguard our own species. Nevertheless it is man who normally is subject to the highest concentration of pollutants, so that he can be said to be acting as a "guinea-pig" for wild life. This is the reason why the countryside is not more seriously damaged though there is no excuse for complacency, or for underestimating the damage in urban and industrial areas. Suspicions that sulphur dioxide and other substances may be more harmful than is at present accepted may make us even stricter in our controls. Pure air in an industrial civilisation is expensive, but it is possible. Already our larger chemical manufacturers have spent millions of pounds on reducing air pollution. There are even vested interests at work. I saw recently a paper entitled "Long-range economic effects of the 1964 Clean Air Act"; I expected it to deal with improved agriculture and health. In fact it foretold up to fifty per cent increases in sales for equipment to control air pollution! Let us hope this target is reached.

Nevertheless we find it difficult to deal with one form of atmospheric pollution, that is with unpleasant smells. Man is not considered to have his olfactory senses particularly well developed as compared with some other mammals, yet he can detect the presence of many odours at a concentration which cannot easily be confirmed by methods of chemical analysis. Anyone who has suffered from smells from farmyards, manure spreading, piggeries or even from chemical factories knows how difficult it is to have such a nuisance abated. He will probably be told that he will soon "get used to it," and is only certain of more serious consideration when poisonous substances can be detected in amounts which can be shown to be dangerous. The difficulties of stopping intermittent smells being given out from farms or factories are such, and the legal costs which may be incurred without the certainty of success (and then with the prospect of paying the legal expenses of the persons causing the smell) are so great, that many people sell their houses at a loss (hoping that prospective buyers will call when the wind is in the right direction) and move away to another district.

If other mammals have a so much keener sense of smell, they must be even more distressed, perhaps by odours to which we do not object or which we cannot detect. I know of no proof of animals leaving an area because of a smell which is also not toxic, but it seems probable that this sometimes happens. On the other hand the stench in the dens of some carnivores suggests that they are even more tolerant than man of some types of smell.

There is one important point about air pollution which is not always remembered. People complain, usually quite wrongly, that polluted air is short of oxygen, and they believe that they inhale more of this vital gas in the country or on the top of a mountain than when in a town. In fact there is little change in the amount of oxygen in the air even in the stuffiest room; there is certainly more in a crowded lecture room in London than in the rarer, though purer, air at the top of Ben Nevis. Industrial pollution, except for the undiluted exhaust gases from chimneys and engines, hardly reduces the amount of available oxygen. Carbon dioxide, present in pure air in very small quantities (approx. 0·03 per cent) is indeed increased by pollution, but seldom if ever to a concentration which is harmful to animals, and it may even stimulate plant growth. Man's breathing is upset by air containing 7 per cent of CO_2, and 14 per cent breathed for some minutes can be lethal; such levels of pollution have never been recorded except in such enclosed spaces as fermentation chambers in breweries. "Stuffiness" is experienced in crowded rooms, but this is not due to the shortage of oxygen or the amount of carbon dioxide. It is due to very small amounts of organic substances given off by the other occupants of the room ("B.O."), and to shortwave radiation from the walls and people themselves. Many Englishmen – and even more English-women – think a room is stuffy and "polluted" simply because, for once, it is comfortably warm! Polluted air is usually "normal" air, in so far as its content of oxygen, nitrogen and carbon dioxide is concerned, plus the addition of small quantities of added materials. Polluted water, as will be seen in the next chapter, may provide quite different problems.

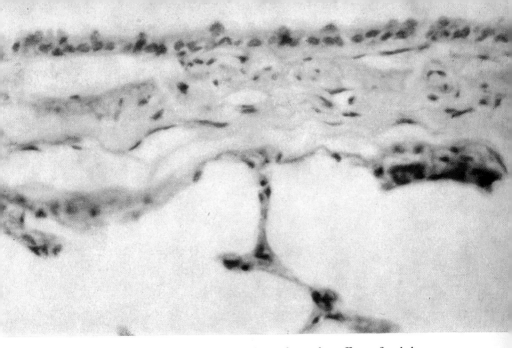

4 Photomicrographs (enlarged × 475) to show the effect of sulphur dioxide on mammalian lung tissue. *Above*, the intrapulmonary airway or 'breathing surface' of a rat. The specimen has been stained (periodic acid Schiff) for mucin, which shows purple. This rat has breathed clean air. *Below*, the same, from a rat which for 3 hours a day over 3 weeks has breathed air polluted with 400 p.p.m. of sulphur dioxide. The photograph shows strikingly how the lung reacts to the SO_2 by greatly increased goblet cells which are congested with mucin. Bronchitis follows

CHAPTER THREE

WATER POLLUTION

As was seen in the last chapter, air pollution has proved difficult to study, and many conflicting results have been obtained. Water pollution seems to have provided a more satisfactory topic for investigation. This is not because the subject is simpler; in fact in some ways it is even more complicated. Air almost always contains sufficient oxygen to sustain life, and "pollution" only means adding a lesser or greater amount of some foreign substance to an otherwise wholesome atmosphere. Water, on the other hand, may be greatly depleted of oxygen so that it cannot sustain most kinds of life, or it may have various substances added, so that animals and plants are poisoned. Some waters have both reduced amounts of oxygen *and* appreciable amounts of poisons in solution; the inter-relations between these factors, and their effects on aquatic animals and plants, may be very complicated. The advantage of water over air, from the point of view of the research worker, is probably that it forms a definite and restricted environment, from which animals cannot easily move. It is therefore possible to study the long-term effects of pollution under fairly constant conditions, and there is no difficulty in demonstrating the serious effects which pollution can produce. It may also be suggested that water is a marketable commodity of considerable economic value, while air is, theoretically at least, "free." So although fresh water covers under one per cent of the surface area of Britain, there are probably more scientists studying its pollution than there are investigating air which covers one hundred per cent of the globe, land and water alike.

Several excellent books on water pollution, and books on fresh-water dealing authoritatively on aspects of the subject, have been published in recent years. These include *The Biology of Polluted*

45

Water by H. B. N. Hynes, *Fish and River Pollution* by J. R. Erichsen Jones, the New Naturalist *Life in Lakes and Rivers* by T. T. Macan and E. B. Worthington, and *Freshwater Ecology* by T. T. Macan. The existence of this extensive and easily obtainable literature has enabled me to make this chapter much shorter than would otherwise be desirable, and to deal with the effects of pollution in a rather different way than would have otherwise been possible.

Man's requirements regarding water are different from those of "wild life" generally. Man demands what he describes as "pure" water; what he really means is "safe" water. This must contain only a minimum amount of salts, and must be free from those bacteria, protozoa and arthropods which might develop in his body and cause disease. Man thus deliberately interrupts the life-cycle of many other forms of life by the various methods of purification which he uses. He is less concerned with the oxygen content of the water than are the fish and insects which live in it. These efforts to produce a pure water supply for city dwellers can even be thought of as a form of "pollution," in that water catchments alter, and sometimes sterilise, large areas of the countryside. The conflict between Manchester Corporation and many naturalists and others over the fate of much of the English Lake District illustrates this point.

On the other hand, water which, for public health reasons, is considered to be "grossly contaminated" by sewage, may still be, from the biological point of view, a healthy and desirable environment for many animals. But by the deliberate discharge of his domestic and industrial wastes man most greatly affects streams and lakes, and so alters the whole composition of their flora and fauna.

Natural waters may not only be "impure" from man's point of view because of the parasites they harbour; they may contain many substances, even poisons, without any human intervention. Quite high concentrations, sufficient to poison some fish and many insects, of lead and copper are found in waters which percolate through strata rich in these metals. Streams running through forests, particularly pine forests, may be contaminated with large amounts of organic matter, and the results may be quite similar to those arising from domestic pollution. As a rule a special flora and fauna is found, consisting of plants and animals adapted to such conditions, in these impure waters. Human pollution usually happens so quickly that impoverishment occurs, often without time to allow the introduction of many of these special types of organism.

Primitive man did not seriously harm the aquatic environment. He often lived beside rivers and lakes, and his waste products must have entered the water, but in insufficient quantities to have adverse effects on the flora and fauna. In fact excrement entering the water in this way no doubt contributed to its nutritive value, and the substances it contained entered into the normal cycles. In some of the less developed and less densely populated areas of tropical Africa we can see a similar situation to-day. The streams and ponds are full of healthy fish; the human beings have a rich internal fauna of parasite worms which pass part of their lives in the water, inside small crustaceans or fish. Man in this way contributes to the richness of wild life in his environment.

When man came to live in towns and cities, however, his increasing numbers had a very different effect. Sewage continued to be poured into the rivers, but the quantities were so great that most unpleasant results were obtained. By the middle of the nineteenth century the Thames, and many other major British rivers, had become open sewers. There are many accounts in the literature. I myself like the account of the Reverend Benjamin Armstrong, from his diary:

> "July 10th, 1855. Took the children by boat from Vauxhall Bridge to show them the great buildings. Fortunately the Queen and Royal Princes drove by. The ride on the water was refreshing except for the stench. What a pity that this noble river should be made a common sewer."

Practically every other river was treated similarly. Even the Cam flowing through the Backs at Cambridge was in this way abused, as is illustrated by the (perhaps apocryphal) story of Queen Victoria's conversation with the Master of Trinity when she looked over the bridge. "What," she asked, "are all those pieces of paper in the water?" The Master promptly replied, "Those, Your Majesty, are notices saying that bathing is forbidden."

The results of all this untreated, or "raw," sewage, vary greatly, depending on the volume of water and the amount of organic matter. As indicated above, small amounts of raw sewage may be actually beneficial to most forms of aquatic life. To-day in some rivers, including the Bedfordshire Ouse, the comparatively small number of boats present discharge the contents of their water closets straight into the water. This does not cause noticeable offence. In some parts of the Norfolk Broads it does, for there are many boats producing much more sewage and this is dangerous. In really crowded rivers, such as the Thames, such disposal methods are not allowed.

Sewage, in quantities which are large enough to have a biological effect, acts in different ways depending on the temperature, the nature of the water and various other factors. The most important biological effect arises from its breakdown by bacteria; this requires oxygen, and as a result the water tends to become deoxygenated, and so less suitable to support most other forms of life. Almost all pollution of water with organic matter, be it sewage, effluents from factories (particularly food factories and dairies) or sawdust and similar wood waste, has this sort of effect. Organic pollution is usually measured by the "biochemical oxygen demand test" (B.O.D.). Experience has confirmed the value of this test, in which a sample of contaminated water is incubated, in the dark, at 20°C. for five days in a closed container containing a known amount of oxygen in solution; the amount of oxygen taken up by the sample is a measure of its B.O.D. Where this is high, and where the diluting water is not present in large amounts, trouble is likely to occur.

It is not generally realised how little oxygen is present, dissolved, in any sample even of "pure" water. A litre of water, at 5°C., in free contact with the atmosphere, only contains about 9 cc. of oxygen, weighing 13 mgs. As the temperature rises the oxygen content falls, so that at 20°C. it is only about two-thirds

5 *Above*, Glasgow, including part of the Gorbals area which is now being cleared. *Below*, Oldham, Lancashire

6 *Above*, the River Trent, just below Nottingham. Purified sewage effluent is discharged just above here, and detergents which pass the filter beds are whipped to foam when the water goes over the weir. Some coarse fish are caught even under these conditions. *Below*, detergent foam blown from the River Aire into Castleford, Yorkshire. Eggs of parasitic worms have been spread in this way

the level at 5°C. As the rate of metabolism of cold-blooded animals may treble with such a rise in temperature, an oxygen shortage is easily produced. Air, even polluted air, is a much richer source of oxygen. A litre of air contains about 210 cc. of oxygen, weighing approximately 300 mgs., i.e. over twenty times as much as is found in the same volume of well-oxygenated water. This may help to explain why some chemicals are toxic in very low doses when dissolved in water; an aquatic animal to breathe must make intimate contact with an immensely large volume of water in order to obtain enough oxygen.

Oxygen reaches the water in two main ways. First, it dissolves at the surface from the atmosphere. Still water takes up oxygen slowly, turbulent water rushing over falls takes it up much more rapidly, for this often submerges bubbles which act as does bubbling air through a domestic aquarium. This type of solution will rarely raise the oxygen level above saturation. The second source of oxygen in water is from photosynthesis. Where there are many green plants present, during the hours of daylight the water may often become supersaturated with oxygen. Unfortunately after dark photosynthesis stops and the plants continue to respire and so actually reduce the amount of oxygen in solution. Therefore during a twenty-four-hour period some waters have a range of oxygen levels which varies enormously, from practically nil around dawn to a very high volume in the early afternoon. Many animals are adapted to life under these conditions. Some biologists have not realised that they exist, and have given too much importance to single measurements of oxygen level in samples of water, not realising that in a few hours far more or far less of the gas may be available.

The capacity of organic pollution to deoxygenate water is enormous. The sewage produced by a single human being gives rise to a daily oxygen demand of 115 gms. (¼ lb.). This represents the total amount of oxygen dissolved in 10,000 litres (over 2,000 gallons) if the water is saturated. In most rivers where sewage is discharged the water, before contamination, is usually far from saturation, so an even greater volume may be affected. Some industrial wastes have much greater effects. For instance it has been calculated that the oxygen demand created by the manufacture of a ton of strawboard corresponds to the sewage output of 1,690 persons, so it could deoxygenate some 17,000,000 litres (nearly 4,000,000 gallons) of oxygen-saturated water daily. These figures are somewhat academic, as they do not allow for the considerable amount of oxygen which dissolves into moving water from the atmosphere. Were it not for this important factor almost any river contaminated with any appreciable amount of organic matter would remain completely deoxygenated; deep lakes, with little water movement, become "purified" much more slowly, and severe pollution can have permanent effects.

There is little doubt that the Thames, formerly an excellent salmon river, reached a peak of pollution, and complete deoxygenation, during the nineteenth century. It was almost entirely due to untreated sewage produced by the human population that this disgusting condition was produced. This is not surprising. The flow of the river may be as low as 200,000,000 gallons a day. The water entering the London area is already depleted of oxygen, and as it is slow-moving only relatively small amounts of further oxygen go into solution. The sewage from a population of 100,000 people would, if the water were originally saturated and if no oxygen were added (and these two factors tend to cancel out), produce complete deoxygenation. It is no wonder that much of the sewage remained undecomposed for days, carried backwards and forwards through the city by the ebbing and flowing tide. Notwithstanding the increased population of to-day the situation, through improved methods of sewage treatment, is in fact considerably improved, at least from the aesthetic, and hygienic, point of view, but the water is still frequently completely or almost completely devoid of oxygen and the fauna and flora are of the kind resistant to such conditions. Pollution is now due not only to (treated) sewage effluent, but also to a great deal of industrial waste, which presents many problems mentioned below.

Many methods have been suggested for dealing with sewage. Ideally it should be returned to the land as fertiliser; if all the salts which we pour down the drains and, eventually, into the sea could be recovered, they would replace the greater part of our imports of chemical fertilisers and might replace them in a more desirable form. Various methods of composting sewage have been devised, and successfully adopted in a few places. In China agriculture in many areas depends on the use of human excreta as manure. The main difficulty is that unless carefully done, the composting process may not kill parasitic worms and other pathogenic organisms, and the compost may be a danger to health. Nevertheless I think that eventually these problems may be solved to the benefit of our rivers and our agriculture.

At one time "sewage farms" were commonly developed. The raw effluent was run into channels and allowed to percolate into the ground. Excellent vegetables were grown on ridges between the channels. Where large areas of well drained soil were available, with no rapid percolation into the water supplies, this was a reasonably safe method, and the material was broken down by the soil bacteria in a fairly short time. An optimum addition of sewage gave maximum fertility and no serious pollution, though parasitic worms and pathogenic bacteria often fouled the vegetables which therefore needed careful cooking. However, there is an upper limit to the amount of material which can be treated in any area as over-treatment overwhelms the bacterial fauna and disgusting conditions result. As suitable ground is becoming less easily available, this method has been largely abandoned.

To-day most urban waste is dealt with before being discharged into rivers, though quite a lot of raw sewage is still run directly into the sea and into tidal estuaries. This latter procedure has in recent years been the subject of much justifiable criticism, as it has been a cause of severe health hazards as well as aesthetic unpleasantness; nevertheless it has probably contributed to the richness of the flora and fauna on the shore near to several popular seaside resorts. The usual methods of sewage treatment depend essentially on oxidation by aerobic organisms. The most widely used system includes filtration through trickling filters, which are the circular structures seen in most sewage works. They are made of clinker or broken stones, and the fluid trickles slowly through them, leaving the interstices full of air. It takes some months for a filter bed to reach its maximum efficiency. It becomes covered with many different micro-organisms which feed on, and so remove, most of the organic matter. The filter is prevented from quickly becoming clogged by their growth because insect larvae and worms also develop in large numbers and feed on the micro-organisms. Another system of sewage treatment is the active sludge process. In this the sewage is run into tanks. These are inoculated with the sludge from a previous batch (to make sure the correct micro-organisms are present) and the whole is kept stirred to ensure aeration. The organic matter is broken down as in the filters. A clear effluent, and "sewage sludge" which is dried and may be sold as a fertiliser, is produced.

These methods of sewage treatment, supplemented by filtration through sand in some cases, are remarkably successful. The greater part of the flow of some of our rivers is in fact treated sewage effluent. It is sometimes said that the water of the Thames when it reaches London has been drunk and passed through different sewage works at least five times. The result is, from the point of view of man, very much an improvement on the conditions which obtained a hundred years ago. There are, however, disadvantages in treated sewage effluent as compared with moderate doses of raw sewage from the point of view of some forms of life. Raw sewage is oxidised rapidly, but its breakdown products become gradually available over a period of several days or even longer. In even the most slowly moving river this means that they are diluted and spread over a considerable area. In treated effluent many salts, not themselves poisonous, are present, and are immediately available as sources of nutrition for plants including algae. Thus a "clean" sewage effluent can have a more rapid, and in some ways more undesirable, effect on the vegetation than the same amount of untreated sewage. This emphasises the conflict that may arise between the needs of human hygiene and the preservation of natural conditions in streams.

Severe organic pollution with complete deoxygenation of the

water is an obvious and undesirable condition. Life is not completely absent. Many bacteria, some producing poisonous or unpleasant gases like hydrogen sulphide, abound. Some of the insects which actually breathe air at the water surface, such as the rat-tailed maggot *Eristalis*, are quite common, but insects which remain totally submerged and all fish are absent. This condition obtains in much of the Thames estuary, notwithstanding the marked improvement that has taken place in recent years.

In many rivers and streams, organic pollution is intermittent. At times it is severe, and complete or almost complete deoxygenation occurs. At other times the water is comparatively pure and oxygen is present. Such severe pollution will kill all the fish, many of the plants and most of the insects and other invertebrates. Recolonisation when it ceases occurs, but which animals and plants reappear depends on many factors. After severe pollution of a stretch of a river, the remainder of which is unaffected, recolonisation is rapid. Careful sampling has shown that some species of "coarse" fish come back even when the oxygen tension is still quite low. Fish have the obvious advantage of being quick moving and able to progress against all but the fastest currents. Many species of invertebrates cannot move so fast, and plants are dependent on water and air currents, animals and other factors for their distribution. Ecologists can quickly recognise a river which is recovering from a period of pollution.

The usual effect of organic pollution is partial, rather than complete, deoxygenation. This is a very complicated subject and for details readers should refer to the books already mentioned by Hynes and Erichsen Jones, and to the excellent work which is constantly coming from the Water Pollution Research Laboratory. It is important to remember that unless organic pollution is very severe, most bodies of water exhibit self-purification to a greater or lesser extent. Fig. 4 shows the changes in a river below an organic effluent outfall. This illustrates a case of severe pollution, but insufficient to cause complete deoxygenation. A shows how the oxygen level drops and the B.O.D. rises just below the outfall; farther down this process is reversed until the oxygen level is fully restored. B shows the parallel changes in the chemical constitution of the water. C and D show how the micro-organisms and the larger animals fare. The recovery of the "clean water fauna" will depend on recolonisation, if, as appears in this figure, it is totally eliminated just below the out-fall. In some cases the purified river will still be richer in nut-rients than above the point where the effluent entered, and the level of the clean water fauna may be actually enhanced. This indicates how moderate pollution, from a small village, for instance, may have little permanent harmful effect. With the growing population of Britain, however, it seems that other methods of disposal than the rivers will always have to be used if the water is to be kept safe for wild life; if it is only to be drunk by man such high standards are not required!

So far we have mainly considered pollution due to organic waste, and consisting of substances which in themselves, and in small quantities, are harmless or even beneficial to life. Many effluents particularly from industry contain toxic substances. Some of these, including phenols and thiocyanates, are usually broken down by bacteria, particularly if they are mixed and diluted with ordinary sewage which, of course, promotes rich bacterial growth. Many, but by no means all, toxic organic substances are affected in this way, but metallic poisons generally pass through filter beds without loss of toxicity. Some metals are remarkably toxic to certain forms of life. Thus copper is used to keep ponds free from algae, when 0·5 parts per million is often effective. Fish survive just over 1 p.p.m., and 2 p.p.m. is tolerated in human drinking water. Zinc affects certain invertebrates at widely different concentrations, some snails are killed by 0·3 p.p.m., some insects survive 500 p.p.m. Much more work is necessary on the long-term effects of metals at low concentrations, not only on fishes and invertebrates, but on man. Recent findings on the effects of low doses of lead on man are disquieting and more harm may be being done to other forms of life than is usually recognised.

One group of organic pollutants which has received special study is the synthetic detergents. They illustrate how serious

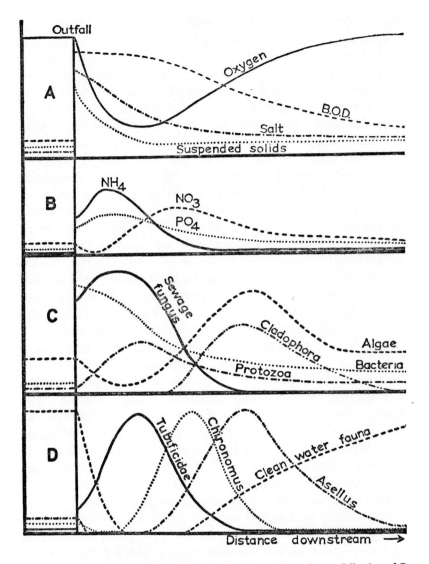

Fig. 4 The effects of an organic effluent on a river below the outfall. A and B physical and chemical changes, C changes in micro-organisms, D changes in larger animals. (From H. B. Hynes.)

damage can be done to amenities and wild life by the unexpected persistence of substances not originally expected to be harmful. The oldest known detergents, the soaps, are made from alkaline salts and certain (weak) fatty acids. The soap which went down the drain was broken down or precipitated in the sewage works and was never thought of again; it did little or no harm. Soaps have the disadvantage that they are relatively insoluble in hard water. Since about 1918 a series of synthetic chemicals has been developed which does not have this disadvantage. Various different chemicals have been successfully used as detergents. The housewife, in her home, has no complaint, and has even come to accept the illusion that they make her washing "whiter than white" when optical whiteners (which have not been shown to have any biological disadvantage nor to make the removal of dirt from clothes more efficient) are included. The trouble has arisen in recent years from the strikingly obvious effect of running the sewage effluent into rivers. As soon as a river has run over even the lowest weir, causing a small amount of turbulence, the detergent has produced enormous quantities of persistent foam which has sometimes caused trouble by blowing in large lumps the size of footballs into crowded streets. The apt name of "detergent swans" has been applied to these aggregations. One interesting point about their occurrence should be noted. The foaming is often worst a long way below the point where the sewage works discharges its effluent into a river. This is because foaming is least in dirty water. It is not until some degree of self-purification of the river has taken place that the maximum foam-production is possible.

The aesthetic damage by detergent foam is obvious. Its biological effects are less easy to determine. Foam blowing from sewage works has been shown to carry pathogenic bacteria and worm eggs, and so is a hazard to human health. Some rivers contain intermittently as much as ten parts per million of detergent without apparently doing a great deal of harm to the flora and fauna, or to the humans who use the river as their water supply. However, these are usually rivers which have to start with a fair degree of pollution and a sparse fauna and flora. It is known that as little as 0·1 p.p.m. of detergent almost halves the rate at which a river takes up oxygen, and so small residues greatly slow down self-purification. Sensitive fish, like trout, are affected by concentrations as low as one part per million, and show symptoms similar to asphyxia. It seems likely that even a very small amount of detergent in a clean upland stream would have a severe biological effect on the sensitive plant and animal life; contamination of such streams from upland farms and cottages must occur.

The economic effect of detergents in sewage works is serious. These substances reduce the efficiency of the filter beds, which must be extended considerably if the effluent is to be maintained at a given standard of purity. Where this fact has not been realised, strongly polluted effluents have sometimes been accidentally discharged into rivers. The reason detergents are persistent, and the foam such a nuisance, is that the molecules are very stable, and that they are not broken down quickly in sewage filters or in the rivers into which the sewage effluent is run. The most troublesome detergents include substances like sodium tetrapropylene benzene sulphonate (TBS); it has a molecule with many branches in its carbon chain (Fig. 5) and this is associated with its resistance to bacteria. Other substances which act as anti-foaming agents, including kerosene, have been added to effluents to prevent foaming; unfortunately they do not remove the detergent, only mask its presence, and may actually aggravate the pollution. Recently entirely new chemicals, as efficient in washing clothes, but much more easily broken down by bacteria, have been introduced. These have straight carbon chains and include Dobane JN sulphonate, and finally sodium alkane sulphonate (SAS) which is at least 99 per cent broken down as it passes through the sewage works. The breakdown products are, as far as is known, non-poisonous. Already some countries, including Western Germany, have made the sale of the so-called "hard" detergents, i.e. those with branched-chain molecules which are so stable, illegal, and some progress with their replacement by the "soft," straight-chain substances has been made in Britain. It therefore seems possible that in a few years this form of pollution may have disappeared. However, there are many useful lessons to be learned from this subject. Detergent pollution might not have been noticed but for the appearance of "swans." Other cases of pollution may go undetected when new substances are used, for instance in industry, so constant monitoring of effluents and the rivers into which they run is obviously necessary.

Other very stable substances are found polluting water, but with more serious results than those resulting from the presence of the relatively non-poisonous detergents. The most serious effects have resulted from the presence of the chlorinated hydrocarbon insecticides, which are both persistent and poisonous to most forms of life, but particularly to fish and aquatic insects. This subject is discussed in detail below (p. 125), when consideration is given to the whole question of environmental pollution by pesticides.

Many rivers to-day are altered by industry, particularly by electric power stations, by having the temperature of the water raised. Where this warming accompanies organic pollution, the effects are greatly increased, as the oxygen level is lowered while the rate of metabolism of the bacteria and other forms of life is increased. In some cases different animals and plants, more adapted to warm conditions, have colonised regions where heated effluents are discharged. This heating of rivers and lakes is something which is likely to increase with growing industrialisation, and it needs to be watched from many points of view, including that of the preservation of wild life.

Rivers may also be cooled, if heat pumps are introduced to warm our cities. These installations remove the heat energy from the water, and pilot plants have reduced the temperature by two or three degrees. This will probably slow down many biological processes, but it is unlikely to have a very great effect on the composition of the flora and fauna.

It is difficult to foresee just what will happen to the lakes, streams and rivers of Britain in the future. So long as sewage and industrial effluents are discharged into our rivers, these will cease to have their "natural" fauna and flora, even if the more unpleasant symptoms of pollution are no longer tolerated. If no

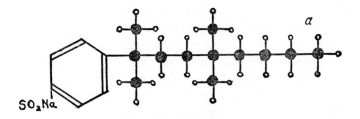

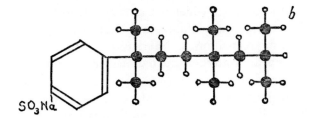

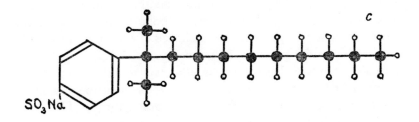

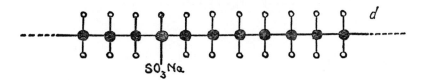

Fig. 5 Chemical structure of "hard" and "soft" detergents.
A: TBS (hard)
B: Dobane PT (hard)
C: Dobane JN sulphonate (soft)
D: Sodium alkane sulphonate (soft)

effluent were to be discharged, under present conditions, many of our river beds would be dry for most of the year. The policy of drawing water supplies from the lowest reaches of the rivers instead of from the cleaner streams above the towns is one which appeals to the conservationist more than it did to most water engineers in the past, though changes in policy now bring these different organisations closer together. Certain river authorities now wish to use their upland reservoirs for storage only, discharging them into the rivers which act as channels to bring the water to the lowland towns. This means that the catchment areas can be used for recreation and agriculture, as pollution is less important when the water, abstracted further down, must be purified anyhow.

If sea-water can be economically desalinated, the pressure on our fresh-water supplies will be eased, and further improvement of their purity will be possible. It may be difficult fully to restore conditions in lakes which have become polluted, and every effort should be made to prevent the discharge of even the "cleanest" effluent into such places. So far comparatively few species of animals or plants have been totally exterminated from fresh water in Britain, and streams which have been cleaned up have usually been recolonised with the appropriate forms. This may not happen in the future unless every effort is made to render our fresh waters not only safe but also pure.

CHAPTER FOUR

RADIATION

Since the first atomic bombs were dropped on Japan in 1945, we have been aware of the dangers of "atomic radiation." Radiation, in sufficient quantities, is clearly dangerous to human and to other forms of life. Man has polluted the earth by releasing radiation and radio-active materials as the result of testing nuclear weapons, by accidental discharges from nuclear power stations, by the waste products from these power stations, by the use of radio-active substances in industry, in research and in medicine and by the use of X-rays in clinical diagnosis and in the treatment of disease. My task here is to assess the danger to man, animals and plants from the pollution that has so far occurred, and to discuss possible future risks from this source.

This is not the place for a detailed discussion of the nature of radiation, but some account is necessary for readers with little background knowledge of the subject. Atomic radiations, which are physically of several different kinds, some consisting of electromagnetic waves and some of bits of atoms moving at high speeds, all have similar chemical and biological effects. These radiations are invisible, and they penetrate living tissues to a greater or lesser extent; some are stopped in the first fraction of an inch of skin, others go deep into the body. All the radiations we are considering here are called "ionising radiations"; this means that they have the property of knocking out electrons from the atoms in the substances they pass through, and producing "ionised atoms" which have great chemical activity. When this occurs in a living cell, the usual effect is for the cell to be damaged. A large dose of radiation, producing many active ions, can cause the cell to die almost immediately. A very small dose may have no noticeable effect, though even the most minute amount of radiation causes *some* change in some part of the cell.

Ionising radiations arise from various sources. Before man made his contribution natural background radiation existed, and is still the most important source in most areas of the world, and it has the greatest effect on its human and other inhabitants. Three-quarters of the background radiation affecting man comes from outside his body; a third of this fraction is due to cosmic rays reaching the earth from outer space, and two-thirds is due to the local radioactivity of many of the rocks. Cosmic ray radiation is partially absorbed by the atmosphere, and is therefore much greater at the top of mountains; at higher altitudes it may increase enormously, and could be a serious hazard to astronauts. Even high-flying birds will receive more cosmic rays than those which remain near the ground. The radio-activity of rocks varies in different parts of the world. Near uranium deposits it may be high, but the differences between different parts of Britain are probably under 50 per cent.

The rest of the natural radiation affecting man comes from radio-active substances within his body. The most important of these is potassium. Potassium is an essential constituent of all living tissues. A tiny fraction, about one part in ten thousand, of this naturally occurring potassium is radio-active, and this produces radiation within the body. Radiation generated within the body in the vicinity of a vital tissue may be very dangerous, compared with similar radiation coming from outside and perhaps being absorbed so as to cause only superficial damage.

Man has contributed to radiation exposure in various ways. First he has concentrated naturally occurring radio-active substances, particularly radium, and used their radiations for medical diagnosis (X-ray examinations and photographs) and for certain types of treatment. More recently he has learned how to produce radiations for scientific research, and to use them for industrial purposes. Finally he has learned to "split the atom" and from this knowledge have stemmed both nuclear weapons and nuclear power stations. A nuclear explosion is accompanied by intense radiation, the effects of which may go farther than those of the blast and heat. Secondly the explosion liberates a quantity of radio-active dust, which goes high into the sky and is slowly deposited as "fall out." Atomic power stations produce radiations which would be dangerous to life nearby if precautions were not successful. The greatest worry relating to these stations is, however, that they produce large amounts of radio-active waste products which could cause great danger if not disposed of properly.

It is necessary to define a unit to measure radiation. Most readers of even popular non-specialist articles find the multiplicity of units and technical terms used confusing, though these cannot be avoided in any detailed discussion of the subject. I shall here confine myself to one unit, the "rad," which can be used as a measure of the amount of radiation reaching a man. If he is exposed to 1,000 rads he will be killed, dying in less than a week. Natural background exposure amounts to 0·1 rads (or 100 millirads) a year; man has evolved under these conditions, so it is generally assumed that this dose is relatively harmless. Radiation exposures between these extremes have different effects. A dose of 50 rads may cause cancer to develop or may damage an embryo *in utero*. A similar dose may cause temporary or permanent sterility if it reaches the gonads.

Even smaller exposures may affect the germ cells and give rise to increased numbers of mutations, that is they cause genetic changes which are inherited. As it is upon mutations that evolution depends, some people wonder why there is so much concern about these effects of radiation. The reason is that we find that the vast majority of mutations are inferior, only a tiny minority are advantageous. In natural selection this minority could be perpetuated. Under the sheltered conditions under which man lives, many of the inferior mutations will be kept alive. The genes carrying harmful mutations may also remain undetected for many generations, so some of the hidden effects of small amounts of radiation may not appear for hundreds of years. Mutations in wild animals and plants may also be caused by radiation. These will probably be quickly eliminated, but may persist, particularly in isolated populations.

Radiation damage differs from most other processes in that its effects are cumulative, and once damage has occurred it is almost irreparable. The effect of 50 rads exposure in one year is very roughly similar to that of 5 rads continued over ten years. Health precautions are therefore usually based on the total dose received during the life of an organism.

The various radio-active substances, which may be absorbed into the tissues of animals and plants, have different properties, particularly in relation to the length of time for which they give out their radiations. This takes place as an emission while the element concerned is transformed into a different element; the process is known as "radio-active decay." Thus uranium eventually turns into lead. The rate of this transformation, and of the production of radiation, differs greatly in the different radio-active isotopes of various elements. We speak of the "half-life" of a radio-active substance, meaning the time in which half of the radiation is lost. Thus the half-life of strontium-90 is twenty-eight years; it gives off half its total radiation in twenty-eight years, three-quarters in fifty-six years, seven-eighths in eighty-four years, and the remaining twelve and a half per cent at a decreasing rate between then and eternity. Other substances have half-lives of fractions of a second or of thousands of years. The atoms which decay very rapidly are usually of little danger because of this instability. Atoms with immensely long half-lives give off proportionately less radiation, so are again not so dangerous unless present in large amounts. Intermediate substances, particularly when, like strontium, they are isotopes of elements normally occurring in living organisms, are the greatest potential menace.

Until recently the only cases of serious damage from radiation were to people dealing with radio-active materials. Most of the early radiologists suffered severe irreversible damage, and factory workers using luminous paint absorbed doses which eventually killed many of them. We have only gradually come to realise the potential danger of radiation, and to-day only permit about one-hundredth of the dose which was considered "safe" thirty years ago. Some scientists think we still allow too high a dose,

7 *Above*, Chelsea creek, London, containing no fish and almost no other forms of life.

Left, flotsam in the River Thames below London

8 *Above*, experimentally-produced radiation damage in an American forest. *Below*, a massive tank of lead, sheltered by concrete a yard thick, for storing radio-active waste. Such tanks may be dumped at the bottom of the ocean

particularly in view of possible genetic effects, though the permitted dose is usually well below the unavoidable natural background level.

Although the most widespread radio-active pollution comes from nuclear explosions, the main immediate effect of an atomic bomb explosion is not due to radiation, but to the blast heat from the explosion itself. The two small prototype bombs dropped at Hiroshima and Nagasaki in August, 1945, reduced large areas of the towns to rubble and killed almost all the inhabitants within a wide radius. About four-fifths of the Japanese casualties are estimated to have been due to the force of the explosion, and only one-fifth due to radiation. Some of the victims near to the explosion were killed almost instantaneously by receiving an enormous dose of radiation, but most attention has been focused on the effects of radiation on people, usually a little farther away, which did not cause instant death. Some appeared at first to have suffered no harm, but died within hours, days or even weeks. Others appeared unharmed for much longer periods, but developed symptoms many years later. The contrast between the effects of conventional weapons, immediately apparent so that the worst effects could be appreciated at once, and radiation which could cause no apparent damage in victims already doomed from its effects, has proved increasingly frightening to many people.

Death, immediately after exposure to a single intense dose of radiation, is only likely in war or as a result of a disastrous failure in an industrial plant, and only small areas of any country are likely to be affected in this way. If mankind becomes involved in a general nuclear war, bombs by their blast and radiation may immediately destroy many of the main centres of population and of industry, but plants and animals in more remote areas are likely to be unaffected at least for a short period.

The wider dangers from nuclear explosions are due to the fall out. All these explosions produce some quantity of radio-active dust. This mostly rises to a height of many thousands of feet, and then slowly descends again to the earth's surface. Different kinds of bombs produce different amounts and kinds of radio-active substances, but all give some fall out. Large dust particles come down quickly. These, the "near fall out," reach the earth's surface soon enough for short-lived isotopes to be a source of danger. Nuclear explosions have thus caused deaths to man and to wild life over areas of thousands of square miles, large areas indeed but only a very small fraction of the globe's surface. Smaller particles take weeks or even years to descend, so that short-lived isotopes decay and only the longer lived ones are important. There is no doubt that delayed fall out has caused and is still causing radio-active contamination all over the world, but the amount up to date is only a small fraction (about one per cent) in comparison with background natural radio-activity. Some believe that strontium-90, which has increased some hundredfold, and which is found in milk and thence in children's bones, where it is in the position to do the most harm, is a serious health hazard. Others state that *at present*, as radiation from fall out is only such a small percentage of unavoidable natural background figure, the danger is negligible. The way strontium-90, and other radio-active substances are concentrated by living organisms is further discussed below in connection with the disposal of power station waste products.

I do not myself think that, except in the vicinity of nuclear test explosions, fall out has caused measurable damage so far to animals and plants, even though a case can be made for the suggestion that some substantial number (at least several thousand) of the 3,000,000,000 inhabitants of the world have developed leukemia or other forms of cancer as a result of this source of radiation, and that some genetic damage, affecting perhaps a very small *percentage* of individuals, but in total a large *number*, has also occurred. Enormously large numbers of plants and animals must also have been affected, but nevertheless, up to the present, man-made radiation can seldom have had important ecological effects. Nevertheless we are all agreed that extraneous radiation should be kept to a minimum, and that testing nuclear weapons should, if possible, be stopped, though some consider possible risks from radiation must be suffered when political expediency dictates.

In a global nuclear war, fall out would eventually wipe out mankind and most other forms of life. Different species of animals and plants differ considerably in their susceptibility to radiation; thus snails are ten times, and the unicellular *Amoebae* one hundred times as resistant as man. The most resistant forms might survive locally and colonise the more heavily radiated areas when these became again inhabitable. Unfortunately none of us would be likely to be there to study a fascinating ecological experiment!

We can only hope that mankind will be wise enough to avoid this form of destruction. Man *has* shown considerable foresight in dealing with radio-active waste products, and has controlled pollution by these substances. Had atomic power stations and factories been developed a hundred years ago, when industrial pollution of a type not tolerated to-day was accepted, the damage to our rivers and to the country generally would have been terrible. However, a great deal of radio-active waste is already being produced, far greater amounts are likely to be produced in the future, and safe disposal of these substances is a serious problem.

It is usually accepted that if toxic wastes can be sufficiently diluted, for instance by being liberated well out to sea, they will do a negligible amount of damage. However, this cannot be done with large amounts of radio-active waste, for the acceptable concentrations, after dilution, are so low. We may take the metal copper to illustrate this. Ordinary non-radio-active copper is moderately poisonous to man, more so to fish and to fungi (it is used to control fungus diseases of plants; see p. 98). Experience has shown that three milligrams of ordinary copper in a litre of water (three parts per million) can be safely tolerated by man in drinking water. The permissible concentration of radio-active copper (Cu-64) is $2 \cdot 5 \times 10^{-11}$, i.e. 0·000000000025 milligrams per litre. This means that if only some 40,000 kilograms (about 40 tons) of the radio-active isotope were uniformly distributed throughout the oceans of the world, all that bulk of water would be unacceptably radio-active. This could in fact not happen, first because Cu-64 decays too rapidly (it has a half-life of only about 13 hours) and secondly because even when deposite   far out to sea effluents are restricted to small fractions of the ocean and are carried, in comparatively concentrated form, by currents. The example nevertheless serves to illustrate the magnitude of the problem.

If radio-active wastes can simply be stored near their source, the products with a short half-life soon cease to be dangerous. Storage is not a simple problem. The stored materials give out a great amount of radiation, and protective walls of impenetrable materials are costly. Nevertheless, most establishments producing waste products have developed processes for concentrating these for easier storage, and keeping them until the level of activity has fallen. The longer-lived radio-isotopes have then to be dealt with. Eventually these may be shot into outer space, where they will orbit until they have decayed into harmlessness, but at present this is not a practical solution.

Some wastes have been dumped deep in disused salt mines. Worked-out coal mines would seem to be useful receptacles, but many are connected with water-bearing strata, and contamination of water supplies would thus be possible. Salt mines are seldom connected to underground waters, but unfortunately there are not many of them. Dumping deep in the earth cannot be a complete solution, as few safe places seem to exist.

In the end most radio-active waste is likely to go into the sea. Some is put in sealed containers, which are then sunk to great depths. Ideally the walls of the containers should contain the radiation. This would be very costly and is not considered necessary if the objects are sunk deep enough so that the radiations cannot penetrate to the surface, and if the containers are strong enough not to disintegrate before the substances have ceased to give out substantial radiations. It is accepted that these dumps may do substantial local damage to fish and other marine organisms, but only over a very limited area. The development of various new methods of containing residues, for instance in an insoluble form in glass-like substances, has made good progress.

It has nevertheless proved impossible not to have to liberate some radio-active wastes freely into the sea. This has only been done when the scientists concerned are satisfied that it will not subject any human being to a dose of more than 0·5 rads a year, and when the exposure of any large number of people will not increase by more than one rad in thirty years (that is, by an amount of radiation approximately a third of the natural background). Governments have agreed to this, and have stated that these limits must not be exceeded whatever the cost.

If all radio-active elements liberated into the sea were dispersed throughout the oceans, and remained so dispersed, the problem would be simpler. World-wide agreement on tolerable levels of radio-activity, combined with a knowledge of the rate of decay of the materials already in the oceans, would enable each country to be given a quota for the next year. Unfortunately the situation is not simple. First, wastes are not uniformly dispersed, but are often carried by currents in a fairly concentrated form to one location. Secondly, some elements are taken up and concentrated by living organisms. This process of concentration has something in common with what happens to certain insecticides (see p. 126), but the two processes are really rather different. All living organisms concentrate the elements of which they are made from their food. In the case of marine organisms many substances are extracted from the water. This process of extraction can be incredibly efficient; some elements are a hundred thousand times as concentrated inside the tissues as outside in sea-water. There is nothing odd in the way radio-active isotopes are affected. They do not actively infiltrate the body, but when they form part of the store of an element in the sea and when this is absorbed and concentrated they are absorbed too.

Safety precautions for the disposal of wastes into the sea insist that no serious risk should exist even allowing for these concentrated effects, and for further concentration when man eats contaminated plants or animals. The discharge from British power stations is thus controlled. Experiments have shown that the greatest possible risk from the effluent of the Windscale power station is through seaweed growing in the vicinity which is harvested and made into "laver bread." This is usually eaten only in small quantities, and the largest amount regularly consumed by some people in South Wales is about three ounces a day. The Windscale effluent is controlled below the level where, under the most favourable set of circumstances, it could contaminate the seaweed so that this gave the greediest laver bread eater an increase in radiation of one-tenth of the natural background to which he was already exposed. Similarly at Winfrith the major danger comes from lobsters. Most people eat only a few ounces of lobster meat a year, if they take as much, but one confirmed lobster eater ate this food daily, and his feeding habits caused the amount of radio-active waste discharged from Winfrith to be reduced by a considerable amount.

The precautions at present enforced for dealing with radio-active wastes seem to many to err on the side of being over-cautious. It is clear that if we spent, proportionately, the same amount of money in controlling pollution from other sources, all our rivers would be pure and the air free from smoke and sulphur. Nevertheless I am sure that we are right to be as careful as we are. Once a radio-active material has been liberated it cannot be recovered, and we still do not know whether concentration, as done by seaweed and lobsters, may not proceed on even greater scale among other organisms, and whether elements we think of as having no biological importance may not be amassed in some cases. No one could have foretold that vanadium, a rare metal present in minute traces in the sea, would be picked up in considerable concentrations by sea squirts. There may well be biological processes as yet undiscovered in which the hazards from radiation are substantial.

The present position can be summed up quite briefly. Those who have worked with radiation and radio-active materials in the past have often been seriously damaged or even killed, and workers to-day take all manner of precautions. Man, plants and animals have suffered ill-effects from radio-active pollution in the vicinity of nuclear explosions. The fall out from these explosions, to date, has probably not done a great deal of damage, but with increased testing of weapons, or a nuclear war, great damage could be done. Nuclear power has done little harm.

Some local pollution of a serious nature has occurred when an accident has damaged a power station, but usually this has been prevented. So far radio-active waste products have had comparatively little effect on wild life in any country. Except in deliberate experiments the effect to date of radiation on wild life in Britain is negligible.

The following table shows the present level of radiation from these sources:

Source	Annual dose in Rads
Natural background	About 0·1000
made up of:	
Cosmic rays	0·0250
From rocks	0·0500
From within the body	0·0250
Fall out	0·0013
Waste disposal	0·0003 to 0·0030

There is, however, no reason for any relaxation of vigilance. We still do not know whether these tiny increases may not be harmful, and without strictly enforced controls far greater pollution would undoubtedly occur.

At the moment, however, it is probably safe to say that in Britain the greatest damage to wild life from nuclear developments has been due to the building of large power stations in remote areas which previously were important habitats of many species of plants and animals; these habitats, their fauna and flora have therefore been destroyed . . . but not by radiation!

CHAPTER FIVE

POLLUTION OF THE OCEAN AND THE SHORE

Two-thirds of the surface of the globe (some 140,000,000 square miles) consists of water. The oceans of the world contain approximately 1,400,000,000,000,000,000 tons of sea-water, which is equivalent to the amount of water in Windermere for every man, woman and child alive in the world of 1966. It is therefore not surprising that a great deal of waste material can be safely dumped into the sea. Unfortunately such dumping can do harm to man and to wild life, usually because the materials are not distributed uniformly throughout the seas, but may tend to remain near their source, or a body of water with a large amount of waste material may keep together and move the pollution from a relatively unobjectionable place, for instance many hundreds of miles from land, to a frequented beach.

Our newspapers in Britain frequently complain that much untreated sewage is piped out into the sea. This happens quite near to some of our holiday resorts, and the sea shore in such places can, under certain weather conditions, become foul and unhygienic. This practice is disgusting to man, and will no doubt be prohibited before long, but it does little harm to wild life. In fact both vegetable and plant life is usually rich near sewers; the unhealthy concentrations of organic matter found in rivers and lakes seldom occur in the so-much-larger sea. Complaints are sometimes made that the sea is polluted with vegetable waste, such as material from canning factories in Lincolnshire. The only effect of this dumping seems to be improved growth of mussels and other animals in the region. Treated sewage effluent also enters the sea, some in the rivers (which, at their mouth, are mainly purified effluent) and some by direct pipes from sewage works. This again will be rapidly

72

diluted, and the nutrient salts will encourage plant and animal growth without contributing to deoxygenation and algal growth as in rivers. The nutrients in sewage are wasted when dumped in the sea, as far as British agriculture is concerned, but they do good rather than harm to marine and sea-shore life. The special case of the disposal of radio-active waste is discussed on p. 68.

The most serious pollution of the sea, and of our beaches, is from crude oil. This has been going on for years; the notorious incident of the "Torrey Canyon," wrecked in March 1967 off the coast of Cornwall, was so serious because of the enormous amount of oil involved. Similar incidents on a smaller scale are only too common. All round Britain patches of black, sticky oil cover our rocks and sand, and make things very unpleasant for holiday-makers. All too often we are treated to the spectacle of sea-birds covered in oil. These may have been killed, and hundreds of corpses are often collected, sometimes already decayed and stinking and a source of serious offence. At other times birds not yet dead are observed; many people try to rescue them and remove the oil. Although some survive with skilful treatment, most die, and ornithologists now think that heavily-oiled birds are best mercifully destroyed. Patches of seaweed and colonies of invertebrates also suffer from oil, though little research has yet been done on this subject. The effects on fish have seldom been observed.

Unpleasant as are these effects of oil, they probably have less effect on bird populations than is commonly imagined. Most birds killed are of common species, some of which (e.g. many gulls) have increased in numbers in recent years, and deaths from oil are negligible in comparison with deaths from natural causes. The gulls removed simply leave more food for others which would have otherwise probably died of starvation. No species seems to be greatly at risk, though the numbers of one or two rare ducks may have been reduced. Local populations of guillemots in some islands of the coast of Scotland may have been harmed, though in others numbers have probably increased. The rather small colonies of guillemots and other auks which have hung on precariously in Cornwall against human pressure may well not survive the losses caused by the "Torrey Canyon." Oil is a killer, but we see here, as with so many other cases, that many forms of wild life can afford to lose large numbers if the important parts of their habitat are not affected. Naturally we all wish to end the scandal of oil pollution; fortunately some progress has been made to that end.

With oil pollution, prevention is much better than any cure at present known. Attempts have been made to clean up fouled beaches, using various detergents to dissolve the oil. This has proved successful though costly, and is worth doing in a frequented resort. Unfortunately in the concentrations required the detergents are poisonous to plants and invertebrates, and some observers think that, from the wild life point of view, the oil, which only harms organisms actually covered, is best left alone, once it has polluted a limited area, as attempts to remove it may do more harm than good. Much more research on the effect of detergents is needed. It may be that the "Torrey Canyon" has provided us with a large scale experiment which will help to solve the problem.

The oil problem is an immense one. Every year some 700,000,000 tons of crude oil are carried as cargo in tankers throughout the world. Until recently tankers have had to clean out their tanks, and the washings, amounting to 0·4 per cent of the cargo, or some 3,000,000 tons, have been discharged into the sea. Reputable concerns have been careful to discharge the oil at least 100 miles from land, and international agreements to ensure this have been promulgated, but not all countries adhere to them as yet. Also oil patches can easily be carried for several hundred miles without being disintegrated.

It is interesting to consider the size of the problem. If all crude oil cargoes were spread evenly over the oceans of the world, some seventeen pounds of oil would cover each acre. In fresh water this would be sufficient to kill most of any mosquitoes in a lake. Even the oil discharged from washed tanks would add over an ounce to every acre of ocean. These amounts seem to give greater effects than do many other substances discharged into the sea (see p. 160). This is because the oil floats; if it were emulsified and mixed throughout the whole volume of the oceans its effects would probably be insignificant (about one part of oil in a million-million parts of water).

Shell International Marine Ltd. and BP Tanker Co. Ltd. have given a lead which should eventually solve the question of pollution by tank washing. They have devised the "Load on Top" system. In this, oily waste from all tanks is collected into one tank, and left to settle. The oil rises to the top; the water can be pumped out from underneath and safely discharged into the sea. A small amount of water remains with the oil, but this does not seriously matter. Crude oil is then pumped in on top when the next cargo is loaded. Presumably this method has the additional advantage that less oil is actually wasted. It is the reverse of the practice in the wine trade, where "bottoms" from a number of casks are put together, the sediment allowed to sink, and a quantity of usable though nondescript wine is obtained! It is a great advance, and when universally adopted it will do much to prevent serious oil pollution. The risk from wrecks, and the discharge of waste oil by irresponsible skippers, will unfortunately remain.

People frequently ask why there has been so much fuss about recent oil pollution, when during the 1939-45 war many tankers containing in total far larger volumes than even the "Torrey Canyon," were torpedoed with, apparently, less effect on wild life. The reason is that most of the torpedoed vessels carried petrol. Much burned immediately, the rest being volatile, evaporated into the atmosphere. Crude oil lasts a very long time; there are several million tons floating on the sea, and some of this will reach our shores, even if it is not augmented, for many years unless improved means for its dispersal can be devised.

CHAPTER SIX

HERBICIDES AND WEED CONTROL

Weeds have been defined as plants growing where man does not want them, and the main work of arable farmers has always been to discourage weeds, without damaging plants which are being grown as crops. To-day we try to control not only agricultural weeds and those growing under similar conditions in gardens, orchards and woodlands, but we also wish to regulate the plants growing alongside roads and railways, and to prevent playgrounds and paths from becoming overgrown. There are, in addition, serious problems concerning plant growth in water, ranging from the control of algae in swimming pools to the prevention of blockage by plant growth of canals and irrigation or drainage ditches.

No one denies the need to control weeds in arable crops, though herbicides drifting or being washed out of crops may do damage to plants and animals at a distance from where they were applied. It is the *method* rather than the intention that may be dangerous. However, there are objections to weed control by any means in roadside verges, rough grazing areas or in water, as in these days of intensive cultivation and when our cities spread to take over more and more of the land, verges and other such places which are not cultivated are becoming increasingly important as reservoirs of wild life. We have therefore a number of different problems concerning the effects of weed control on wild life, depending on where that control is exercised.

Weeds undoubtedly harm arable crops. They compete for soil moisture, for light and for plant nutrients. Some, for instance couch grass, produce exudates which retard the growth of other plants. Others, like thistles in cereal crops, make harvesting more difficult. As a general rule the fewer the weeds, and the cleaner the crop, the greater and the more valuable the yield. Nevertheless some experiments have shown that a moderate infestation —particularly of low-growing weeds, and ones which only appear after the crop has made some growth—may have no effect whatever on yield. There must be many cases where the cost of weed control is greater than the gain in yield which it produces. There are also occasions where a cover of weeds over bare soil prevents erosion: this is very important in tropical countries, but even in Britain weeds can help to hold loose soil. The value of weeds as green manure is more difficult to establish.

I mentioned in chapter 1 that improved weed control by mechanical means reached its peak of efficiency at about the end of the nineteenth century, when labour was still cheap, seeds could be efficiently cleaned, and mechanical cultivation, even though only horse-powered, was effective. Such weed control had a limited effect on wild life. It reduced some of the "traditional" and aesthetically pleasing weeds of corn crops, such as cornflowers, corn cockles and poppies. Inter-row hoeing, by hand or by horse, destroyed some birds' nests, including those of plovers, skylarks and partridges. Weed seeds supplement the diet of many small birds, so weed control reduced this food supply. However, as arable cropping itself has such major effects on wild life of all kinds, mechanical methods of weed control are in themselves relatively unimportant, particularly as the effects are restricted to the cropped area and do nothing to contaminate the surrounding countryside.

Good husbandry, using powered machines, still continues, and plays its part in weed control, but farmers are relying more and more on herbicides. Chemical weedkillers have been known for many years. More than a hundred years ago it was recognised that copper sulphate, sprayed in "Bordeaux mixture" to control fungus diseases of vines (see p. 98) also had the effect of killing some of the broad-leaved weeds growing in the vineyards. Various copper salts were used to control weeds in cereals, but they were costly and not particularly effective. They were seldom used in sufficient quantities to have harmful side effects even on a limited scale, as has been seen to occur with copper fungicides.

The weedkilling properties of other inorganic chemicals were also established many years ago. **Sodium chlorate** and **arsenic** compounds have long been used as "total" weedkillers, but areas so treated could not be planted for months or even years.

Sodium chlorate is considered dangerous to man mainly because of fire risks. Under dry conditions it can explode, and it makes plant residues very inflammable. It is not at all highly poisonous, and although it may be locally applied in heavy doses, as much as 400 lb. to an acre, it is unlikely often to be a serious danger to surrounding vegetation, though it may damage water plants if heavy rain washes it into ponds and rivers. It is persistent, but seldom remains undiluted long enough in water to do prolonged damage.

Arsenical weedkillers, particularly soluble sodium arsenite, were a real danger to man as well as to all forms of wild life, because they are acute poisons whose residues remain toxic for many years. Sodium arsenite was extensively used as a spray to destroy potato haulms, and as a result wide areas of Britain are still contaminated. The soil fauna has no doubt been greatly modified in such cases. Although since 1961 the use of arsenical weedkillers has been officially discouraged, and they have not been used in Britain as haulm destroyers since that date, minor incidents due to arsenic residues are likely to occur for some time to come. Incidentally it should be noted that the relatively insoluble substance lead arsenate, though it is extremely poisonous, is still a permitted insecticide, and some is used on fruit trees. The amount of residual arsenic which is permitted on the fruit, one part per million, and two parts per million of lead, is laid down by law, although some scientists think this amount is too great for safety. The dangers which may arise from the use of arsenical insecticides are discussed in chapter 8.

Sulphuric acid had a limited use to clear land of weeds and was even used in cereal crops, where it killed broad-leaved weeds and ran off the narrow leaves of the cereal plants and did them relatively little harm. There was little residual effect, for the sulphuric acid reacts with substances in soil and plant tissues and is rendered harmless within an hour. Sulphuric acid was to some extent a selective weedkiller, depending on how it adhered to the plants rather than on its lethal properties when contact was made. It was sprayed on, and if it drifted outside the sprayed area it killed other plants and animals which got a substantial dose, but on the whole it can have done little harm to wild life. Sulphuric acid is still sprayed to destroy potato haulms.

The first great break-through in the control of weeds in arable crops began with the exploitation of the herbicidal properties of the **dinitro compounds** (for example **"DNOC"** and **"dinoseb"**) in the 1930s. These substances are contact herbicides, which kill many forms of plant when growing actively. Their great agricultural importance was originally that they killed many of the annual weeds which infested cereal crops without serious damage to the cereal plants themselves. They have little effect on most perennial weeds like couch grass or creeping thistle, for though they kill the part reached by the spray the substance is not translocated in the plant and the underground parts survive to send up further shoots. During the 1939-45 war and immediately afterwards DNOC played a most important part in increasing the yield of food grown in Britain. DNOC has also been used, in high doses, as a general non-selective insecticide, as a means of haulm destruction in potatoes, and as a winter wash for fruit trees (here it is acting as an insecticide and not as a herbicide). As long ago as 1892 DNOC was recommended as an insecticide, and had a limited use as such.

DNOC has done a great deal of damage to many forms of wild life. It is a very poisonous substance, as, to a greater or lesser degree, are all the dinitro or phenolic compounds which have been tested as agricultural chemicals. They are poisonous to mammals either by inhalation or by absorption through the skin. DNOC is so poisonous that at least a thousand human beings could be fatally poisoned by the amount which may be applied to a single acre, and supplies maintained on many farms could each wipe out the whole population of a city the size of Cambridge. DNOC is used at comparatively high concentrations, using as much as ten pounds of the substance to an acre, so there may be more risk of some going astray than when substances used in much lower concentrations are applied.

The use of DNOC is only permitted in Britain if the users wear protective clothing, which itself is very uncomfortable and has caused heat stroke in hot weather. There has been a number of human deaths from DNOC poisoning, and birds, small mammals and all types of insects die from contact with the spray. Fortunately it is rapidly broken down after contact with plants or soil, and does not seem to leave poisonous residues, so that animals or plants killed by DNOC do not seem to damage animals which eat them unless they do so very soon after the spray is applied. There is clearly no sort of concentration of these substances in food chains, so the lethal effects are fully manifest soon after application and delayed environmental contamination does not occur. In the early days of DNOC spraying of cereal crops, a good deal of damage to surrounding vegetation due to the chemical drifting occurred. This was partly because most operators were inexperienced, and the dangers from spray drift had not then been fully recognised. Operators also often failed to recognise how poisonous the substance was to man and to animals. There are reports of children following machinery and being dyed bright yellow. It is surprising that more fatalities did not occur. With proper precautions, DNOC can be used in cereal crops in such a way that little such damage occurs, and so that wild life deaths are reduced and only the animals actually in the sprayed crop are harmed. It is possible to reduce bird deaths by fitting appliances to sprayers which frighten the birds and make them fly away before being wet with the chemicals, but insects and, to a lesser extent, soil animals cannot so easily be scared. It is an undoubted advance that DNOC has so largely been replaced to-day by much less poisonous herbicides.

The next major advance in the control of weeds in cereals was the discovery of the so-called "hormone weed killers" in 1942. These are synthetically produced organic compounds which have certain properties similar to the natural hormones

9 Oil pollution of river and coast and its effect on birds. The fortunate few are rescued by the R.S.P.C.A.: most die. *Above*, oil from a sunken barge in the Thames at Blackfriars, London, killed many swans. Sixty were treated at the R.S.P.C.A. clinic at Putney. *Below*, The Wirral coast, Cheshire. Some ships at sea deliberately discharge oil which washes up on the beach

10 Treatment of railway tracks with 'total' herbicides. *Above,* a special coach which sprays weedkillers when travelling at 40 m.p.h. *Centre,* an overgrown track before treatment; *Below,* the same track later in the season

or growth regulating substances and are more properly described as "auxin type growth regulators." One of the first to be tested, and the substance still used in the majority of cases, is MCPA. This is absorbed by both leaves and roots, and is readily trans-- located in herbaceous plants. Many readers have probably used preparations containing MCPA on their lawns, and have seen the plantains, daisies and dandelions grow rapidly and grotesquely and die; annual broad-leaved weeds are even more rapidly killed. Grass and cereals are not usually affected, partly because the chemical does not adhere to their narrow leaves.

I have mentioned that MCPA is the most widely used herbicide. It and the allied substance 2,4-D are in fact the most widely used pesticides of any kind in Britain; they comprise some two-thirds of the total of all herbicides, insecticides and fungicides applied to our crops. There are in addition some thirty other active chemicals which are used as herbicides and these are marketed in Britain in the form of some 250 different commercial products, each with its particular formulation. These products differ in their efficacy against different types of weeds, some being relatively selective and others more suitable for blanket treatment of a wide range of weeds. Some specially important cases are discussed below.

MCPA and 2,4-D are often described as being "non-poisonous." This is not strictly true, for almost every chemical – even common salt – is a poison if taken in large enough quantities. MCPA is a good deal more poisonous than common salt, but much less toxic than DNOC. It would probably require about a quarter of a pound of the chemical to be swallowed to cause death in man; as it would be almost impossible to ingest this amount, the risk is slight, and in fact I do not think that there have been any reports of human deaths from MCPA notwithstanding the vast amount of the chemical which is used every year. Other mammals and birds are seldom if ever fatally poisoned, even if they are drenched with the spray, and those which enter the crop soon after it is treated are unlikely to be harmed. There have been suggestions that fish may be rather more susceptible, and that run off from fields into rivers may damage them, but this does not seem likely to occur often or to present a serious problem. MCPA and 2,4-D are not very persistent chemicals. They break down in the soil within a few weeks of application, and the residues which are left do not possess herbicidal properties. Within two months of even a heavy application a new crop can be planted and suffers no apparent ill-effects, and normal growth and good yields are obtained. It has been suggested that residues harmful to some forms of animal and plant life may persist for longer periods, but there seems to be no reliable evidence to support this, and a good deal which does not.

The real risk to wild life from MCPA is that it is an efficient herbicide which affects almost all broad-leaved plants, and therefore also all the animal species dependent on them. Spray drift from a crop either due to careless application or to wind can do a lot of damage, and hedges, gardens, woodland and even other farm crops are not infrequently affected, so that those responsible for spraying usually take out insurance policies to cover these risks. However, although more and more MCPA is used each year, operators are becoming more skilful and careful, and serious incidents are becoming fewer.

Provided that MCPA and similar herbicides are properly applied to a cereal crop, the effects are in general similar to those of good husbandry in the days of cheap labour, i.e. the weeds are eliminated. I do not myself believe that this, the greatest use of any agricultural chemical in Britain to-day, is seriously harmful to wild life. Not all conservationists would agree. They regret the decline in the number of the more aesthetically pleasing of the weeds of corn crops, such as the corn cockle, even though this is poisonous and may make batches of flour unsaleable for human food. There may be a case for a "museum" farm where bad husbandry is practised to allow weed species to perpetuate themselves, but it is obviously unwise to try to persuade farmers to encourage weeds in their crops. Farmers who consider that their game, particularly partridges, are more important than their crops may be persuaded to allow annual weeds, whose seeds may add to the food supply of these birds, to flourish at the expense of the wheat or barley, but not where the shooting rights are separated from the tenancy. I believe that conservation must be encouraged outside cereal crops, which are not generally important in this connection.

One particular problem of interest both to farmers and to conservationists arises from the repeated application, year after year, of a herbicide to the same area of ground. Many have feared that this might affect the soil flora and fauna, and ultimately this might adversely influence fertility. Unfortunately this problem has not been very thoroughly investigated, but what results have been obtained are reasonably encouraging. Even where herbicides have been used continuously over twelve years no significant differences in the population of soil bacteria and of microarthropods have been found. Changes in the number of worms are attributable to cultivation rather than to any toxic effects. However, I think that more attention should be paid to possible side effects of repeated herbicide application, even if only to show that these do not occur.

I have already mentioned that different herbicides control different weeds, and not infrequently MCPA will give excellent control of most species but will allow some, including serious pests like wild oats and couch grass, to increase. This makes it necessary to modify the cultivation and to use different herbicides. Fortunately relatively non-toxic herbicides are being developed to treat these difficult weeds, so the risk to wild life may not be increased. The problem is an important one to the farmer, but as these effects are restricted to his fields wild life may not be affected. The main danger is that failure to control weeds may encourage larger doses to be used without adequate precautions instead of careful use of the most efficient (and least toxic) substance.

There is always one risk from the repeated use of any pesticide, and that is that resistant strains of the pest may appear. This has been reported many times with insects, where strains which cannot any longer be controlled by certain insecticides are well known (see p. 131). So far similar effects with weeds and herbicides have not been observed in Britain, but a few cases have been reported in other countries. Thus a common weed in sugar plantations, *Erechtites hieracifolia*, can no longer be controlled by 2,4-D, and we must expect more examples of this phenomenon each year. These resistant strains may have other properties which encourage their spread outside the cropped areas, and so important and perhaps harmful ecological effects are possible. For this reason it is perhaps unwise to be dogmatic about the possible long-term effects of even the most apparently harmless chemical like MCPA.

So far we have been concerned mainly with weed control in arable crops, a process that all agree is necessary, even if some would like to protect several of the more attractive weeds. When we come to consider roadside verges, the problem is more complicated. I have already mentioned that with industrial and urban development, the spread of mechanised agriculture and the disappearance of our forests and woodland, processes which are so graphically described by Sir Dudley Stamp in his *Man and the Land*, roadside verges and hedgerows have become increasingly important sanctuaries for wild life. However, it must be realised that the apparently natural roadside verge is an entirely artificial and man-made habitat. The truth of this statement is easily demonstrated by a visit to any area of heavy clay soil where broad stretches of verge have been left uncut and ungrazed for a few years. The hedge appears to have marched out and become a wide tangled thicket, no doubt a valuable habitat for many birds and mammals, but very different from the "natural garden" which is so greatly admired. If our verges are not to become first thickets and then, in many cases at least, some type of woodland strip, constant attention is needed.

It is the duty of the Highway Authorities (usually the County Councils) to maintain the roads including the verges. The vegetation must be prevented from encroaching on the road, and must be kept short, particularly on bends, to allow good visibility to vehicles. When labour was cheap the grass was cut by hand, and then it was often removed as hay. In some areas tethered goats or cows grazed, and cattle or sheep being moved about the country foraged as they went. This gentle cutting and grazing produced rough grass containing a multitude of flowering plants, generally backed by a hedgerow, the whole forming a fascinating pattern of ecological interest and beauty.

In recent years labour costs have risen, and workers have become scarce. Most farming processes have been mechanised, and so has roadside grass cutting. First ordinary farm mowing machines were used; these could only cut smooth and level areas. More versatile machines have been developed, able to work on rougher ground. The trouble with these is that many areas are too rough even for such tools, and also that this cutting has to be repeated several times in the season. The cost of cutting is considerable, but that of removing the cut vegetation, which is usually necessary, is even greater. The cut vegetation can sometimes be sold as hay, but is usually too weedy and polluted by dust from passing traffic to be worth the cost of collection. If left uncollected the cut material may dry and increase the danger of fire, or in wet weather it may kill underlying plants and produce unsightly dead patches. Mechanical cutting has generally been accepted by conservationists and naturalists, partly perhaps because it is not too efficient and has a minimum effect on the rough areas which the machinery cannot easily reach.

Many County Surveyors thought that the discovery of selective weedkillers had solved their problem. They wished to retain grass verges on country roads for aesthetic reasons and to prevent erosion, but they wanted to get rid of tall plants and to prevent scrubby growth. They hoped that herbicides could be applied more cheaply than the vegetation could be cut, and that the effects would be longer-lasting. The first attempts were most unfortunate. DNOC applied in summer turned the verges into disgusting areas of dead and dying plants, the spray more often than not drifted into the hedges and damaged them, and many birds, mammals and insects died leaving their corpses for all to see. This process was not continued on any large scale for long, though selective spraying of patches of particularly "objectionable" weeds has occurred. Then MCPA and 2,4-D were used. These were believed to be non-poisonous to wild life (i.e. birds, mammals, insects) and were also considerably less expensive than the more poisonous substances. Applications were again made in summer, when many of the more attractive plants were coming up to flower. The immediate result, though perhaps not quite so hideous as that produced by DNOC, was quite horrifying, for the roadsides were soon covered with the twisted and deformed plants where attractive flowers had been expected. The outcry continued, and such spraying has only continued in a minority of counties, though keen but inexperienced individuals may transgress the gentleman's agreement to forgo this practice except on trunk roads and within ten feet of the carriageway.

Herbicides on roadside verges have the immediate effects noted above. They also have long-term effects on the constitution of the flora (and therefore of the fauna dependent upon it). There has been surprisingly little scientific work on this long-term effect, the most important exceptions being some interesting experiments which have continued for some twelve years on Akeman Street in Gloucestershire. The botanical side of this work has been described carefully but the effects on animals do not yet seem to have been assessed.

Repeated use of herbicides, particularly of 2,4-D plus **maleic hydrazide**, produces a grass sward which does not need frequent cutting and which contains few tall plants like cow parsley and hogweed. In general broad-leaved plants become scarcer, but in early spring the treated plots contain quite a number of flowers which appear before the chemicals are applied. If the spraying is carefully timed, the distressing effects of the dead and distorted plants associated with earlier experiments are avoided, partly because many of the susceptible plants are no longer present. Most of the herbicides used in this work are not highly poisonous, and it is unlikely that many forms of animal life are killed directly by their use, though maleic hydrazide could, under some circumstances, be dangerous. This has encouraged the statement that they are "harmless to wild life," which ignores their effect on plants and on the animal species dependent on affected plants. For this reason most conservationists are still rightly suspicious of this method.

However, we must face the fact that the verges must be prevented from becoming overgrown, and the preferred method of hand cutting is unlikely to be reintroduced, except perhaps in restricted areas of biological importance where voluntary labour may be available. Many authorities are now using flail cutters. These are generally preferred to chemicals by naturalists, though we still know little of the biological effect. Flail cutters reduce the vegetation to a mulch which does little harm and need not be collected. They also kill any animals which get in their way, and although most birds and mammals make their escape, some do not and corpses can usually be found after a machine has passed. Most insects on the vegetation which is cut are also killed; they generally survive hand cutting or ordinary machine mowing. The vegetation of an area regularly cut by the flail method will be different from that resulting from hand cutting, and we do not yet know whether the results will be preferred by naturalists. From their point of view the main advantage over herbicide spraying is that odd corners particularly with rough ground are likely to be left by the flail, whereas a spray could easily be squirted wherever the operator wishes; that is to say the flail has the advantage of its comparative inefficiency!

Incidentally, the disappearance of rabbits from many areas where they were common, burrowing into the hedge bottoms and grazing on the verges, has been accompanied by an increased growth of scrub and the need for more cutting. Weedkillers carefully used here may actually help to preserve the herbaceous sward desired by the highway authorities and by naturalists. Flail cutters are, however, usually capable of preventing the scrub from developing.

Total and permanent destruction of all vegetation is sometimes required, for instance on railway tracks, on garden paths and around farm buildings. Mechanical methods such as hoeing were the earliest and are still common. Destruction may be better accomplished by burning, either of dead grass on the surface, or more efficiently using a flame gun. Burning can kill germinating seedlings, and the parts of plants above the surface, but even with the largest flame gun it is almost impossible for the heat to penetrate deep enough to kill underground seeds of perennials, or weed seeds that are buried at all deeply. The difficulty of controlling a perennial weed like creeping thistle is easily demonstrated at the site of a bonfire on infested ground. Though such a fire may have burned hotly for days, it is not uncommon to see thistle shoots appear in the middle of a burned area of several square yards, within a few weeks. This cannot be due to inward growth from unaffected areas; the rhizomes immediately under the fire must have survived. Burning and hoeing has the disadvantage that the soil is not rendered "immune" to further weed attack: where crops are to be grown after the weeds are destroyed this is an obvious advantage, but some lasting treatment is desired on other areas.

Herbicides have proved very useful for total weed control. Sodium chlorate, mentioned earlier in this chapter, is still widely used, sometimes in doses of more than a hundredweight to the acre. The risk of fire is reduced by including other substances, for instance calcium chloride, in the preparation. No plants will grow for at least six months after a heavy dressing of sodium chlorate, but when it is finally washed away it has no serious after effects. Even in high doses there is remarkably little effect on the soil fauna. Borax is also used as a total weedkiller. It is rather more slow-acting than sodium chlorate, and also more persistent. It does not seem to have a great effect on the soil fauna, nor to other forms of wild life.

In recent years synthetic organic herbicides have been increasingly used in areas where all plants must be eliminated. These substances differ in their properties. **Simazine** and **Monuron** are long-lasting and may prevent re-growth for as long as a year. Obviously if they are not used carefully, they may sterilise land outside the area scheduled for treatment. **Dalapon** ceases to affect growth within six to eight weeks after application, and **Paraquat** has almost no effect within hours of reaching the soil. None of these substances, whether persistent as a herbicide or not, seem seriously toxic to animal life, though few experiments on the invertebrate soil fauna have yet been made. If plant growth is prevented for long periods, phytophagous animals will obviously be eliminated also.

Total and permanent weed control is quite an expensive procedure, and many of the areas treated are of limited extent and are not important wild life habitats. Serious contamination of surrounding land would cost money, and therefore is rare. Grass growing in the cracks in paving-stones or weeds coming up through the asphalt in playgrounds must clearly be controlled. However, some of the weeds which grow around farm buildings are species to which many people have an ambivalent attitude. One such weed is the nettle. Few gardeners, even those who are keen naturalists, wish to allow nettles on their own ground, yet the same people grow buddleia bushes largely to encourage peacock, red admiral and tortoiseshell butterflies which have nettles as their larval food plant. Nettles are also important reservoirs of ladybirds and other predatory insects which play an important, though often unrecognised, part in the biological control of pest insects. In this case the nettles harbour overwintering aphids and other insects on which the predators subsist during the time the pest species (e.g. aphids on beans) are not abundant. Nettles grow abundantly under many conditions but they are easily controlled by herbicides and by mechanical methods, and this therefore reduces the population of beautiful and of beneficial insects. Nettles disappear from roadside verges when these are regularly mown, even if no herbicides are used. It is common practice for patches which appear in pasture or along ditches to be treated by spot spraying with herbicides. They often continue with least interference in certain types of woodland, but here, unfortunately, infestation with the caterpillars of the more attractive butterflies is uncommon.

Permanent grass was a traditional feature of British agriculture. To-day many farms in lowland Britain have little permanent grass left, and rely on leys where the grass forms a stage in the rotation of the crops grown, and is ploughed up after two, three or four years. Most leys grown after cereal crops kept clean by the use of herbicides contain relatively few weeds and require no further treatment. Short-term leys are generally of little importance to wild life, as they contain few species except the commercial varieties of plant that the farmer has sown. Even the soil fauna is comparatively poor, due to the repeated cultivations accompanying the arable crops, and the short period under grass gives little time for many of the animal populations to regenerate. Thus it has been shown that the earthworms *Allolobophora longa* and *A. nocturna*, which make the majority of surface casts, and produce the rich stone-free layer of "vegetable mould" which Charles Darwin recognised, take up to seventy years to reach the population found in old grass; ploughing almost exterminates them. Other species of worm may survive ploughing, and *Lumbricus terrestris* flourishes where much organic matter is available notwithstanding the disturbance of cultivation, so that it may be as common in a ley as in old grass. The changes due to herbicides, if they are used, are generally insignificant as compared with the fundamental effects of cultivation on plants and animals.

Where old grass still exists, especially on chalky soil, it is of great interest, because of the wealth of species of plants, and of animals, which exist in it. Some of these species of plants are, unfortunately, "weeds," and reduce the economic value of the grass. Herbicides used on grassland are usually designed to kill the broad-leaved species and leave more or less "pure" grass. This may not always be completely desirable even to the farmer, for some weed species are rich in mineral nutrients which may prove beneficial to stock grazing on them. It has been noticed that under some conditions grazing animals actually select these plants, and some farmers deliberately include weeds which are deep-rooted, including burnet and chicory, in their seed mixture when establishing grass leys.

Herbicides may alter the palatability of some plants. Thus ragwort is ordinarily not eaten by most farm stock, fortunately, because it is poisonous. However, after treatment with herbicides ragwort loses its unpalatability, but not its toxicity, and illness and death to sheep and cattle have been reported.

Chalk grassland is disappearing, not only because it may be ploughed up, but also because it is being overgrown with scrub, and if left alone may soon turn to woodland or forest. This is because rabbits, which kept the grass short and prevented hawthorn and other shrubby plants from becoming established, were so reduced in numbers by myxomatosis. The first effect of the reduction of rabbits was an increase in many of the flowering plants, so that naturalists often welcomed this as an improvement. Soon, however, that is within three or four years, scrub appeared and smothered many of the desirable plants. Work is in progress to see whether herbicides such as 2,4,5-T can be used to prevent this from happening. They are painted on to the stumps of cut shrubs and trees, and prevent regeneration, without, apparently, having any harmful effects on the fauna or on the plants in the surrounding soil. It may well be, therefore, that this will be a case where a herbicide, properly used, serves as a valuable help in conservation of desirable habitats.

Quite a substantial fraction of the remaining permanent grass on lowland farms is relatively unproductive, because it has been invaded with weed grasses which are poor sources of fodder. Most herbicides used on grass can control broad-leaved weeds quite easily, but cannot select between "good" and "bad" species of grass. In the past such grass has been ploughed and reseeded with the desired seed mixture. To-day it is possible to kill all the grass and most of the other plants by one application of a weedkiller such as **Paraquat**, and as the herbicide is rapidly rendered inactive by contact with the soil, immediate reseeding is possible. This process needs to be studied carefully, but preliminary observations suggest that it may often be less harmful to the soil fauna than ploughing or other mechanical methods of cultivation. For instance, the surface-casting earthworms, usually almost eliminated by cultivation, may survive. Here again the herbicide, properly used, may be a help and not a danger to conservation.

In addition to lowland grass, be it temporary ley or permanent pasture, there is a great deal of marginal grass which merges into upland moor. This is grazed by sheep to a greater or lesser extent. Around many upland farms an area of improved grass is found; this has usually been produced by liming, the liberal use of basic slag and other fertilisers, and intensive grazing. Sometimes such areas have been ploughed and reseeded with desirable strains of grass, but often a productive pasture has developed under intensive management. This sort of treatment is only possible over restricted areas. In the past most "rough grazing," which is an increasingly important reservoir of wild life, has remained in that condition because improvement is too difficult and expensive, and many areas even after improvement are not very productive.

Many areas of rough grazing have been ploughed recently, helped by the Government subsidy for that purpose. Some improved pasture has resulted. It is not always realised that "improved" upland pasture is an unnatural plant association, which requires intensive management if it is to be maintained in a productive condition. Many farmers have collected their ploughing-up subsidy and then allowed their land to revert. This has often been unfortunate, for the ploughing has destroyed valuable species which may not reappear and so poor grass of little biological interest remains.

Up to now much marginal grassland has been left alone because it is too rough and steep to be ploughed. Herbicides such as Paraquat may now be used. They are sprayed over the whole area, a process much easier than ploughing or cultivation, and more efficient at killing the existing vegetation. Soon after this spray it is possible to sow the desired grass mixture, and under suitable conditions quite productive grass may result provided proper management is continued. This method has not been in use for a long enough time for its value to be finally assessed, but if it is widely successful the flora and fauna of much of Britain will be greatly altered. This is a typical case of the important effect a pesticide can have on wild life. Paraquat, as we have seen, is a particularly non-toxic substance, which will seldom kill any form of animal life by its direct application. However, as it can be used where ploughing is impracticable, its results can be more serious. The trouble is that it may be used in some areas where the old and interesting vegetation would otherwise have been preserved, and often the new pastures will be neglected and so will not retain their value.

The control of weeds in and around water presents many problems. Even when undesired plants are killed their vegetable matter remains and decays. This can cause pollution and even if toxic substances are not liberated the process of decomposition may absorb most or all of the oxygen in the water with serious effects on fish or other animals (see p. 49, where the effects of water pollution of this nature are discussed).

Green algae produce pond slimes in many situations, from the scum in small garden pools to a covering over the whole surface of a reservoir. This scum can often be controlled by the use of **copper sulphate**. Copper, at a concentration as low as 0·5 parts per million, will often prevent algal growth, with no apparent harm to higher plants or to insects or even fish, and this amount is usually accepted in drinking water by public health authorities. Fish are, however, rather susceptible to copper poisoning, and it is difficult to strike a balance, particularly in alkaline water where the copper may be immobilised, between the amount which will control the weeds without harming other forms of life. There have been many incidents in which too much copper has been added to ponds, with similar results to the careless use of copper fungicides (see p. 99), though mammals and birds do not seem to have been poisoned even when the fish have been killed off.

Rivers and canals can become choked with weeds, either by sedges and reeds growing out from the banks, or by submerged and floating species growing in the channels. In the past the reeds were usually cut by hand, and the edges of the banks were trimmed to prevent the reeds invading free water. The herbicide **Dalapon** has more recently been used for the same purpose, with considerable success and with little apparent damage to animal life. The main trouble, as mentioned above, comes from the decomposition of the dead plant material. However, this can also occur even after mechanical cutting.

Navigable rivers and canals have usually been kept clear by using boats armed with various types of weed cutter. Here again large amounts of plant tissue have to be dealt with. These do not usually die quickly, and so deoxygenation is less serious than after the use of herbicides, but the plants may float downstream and large accumulations may block the channels until the material is dredged out on to the bank. Under these circumstances stinking heaps result, killing the plants on which they lie, and often giving, in succeeding years, patches of nettles or docks instead of the original plants. Less damage to the rivers and their banks obviously results from more frequent cutting, with less accumulation of dead material. At times Dalapon may be used in the water itself, but such large amounts of this not inexpensive chemical are needed that this method of control has limited possibilities. Water pollution by this, and by new herbicides, will always be a possibility, but herbicides are in general less likely to damage aquatic wild life than are the effluents from industry and from sewage plants.

The flora and fauna of all countries is constantly changing, as some species become extinct and new ones are introduced. We tend to take the view that our so-called "native" species are in some way more desirable than are "aliens," forgetting that the process of introduction has been going on over most of Britain since the last ice age, and that many plants and animals which might be expected to be well established have been excluded by what can only be considered as accident. However, it is generally believed that the organisms which have been found here for many thousands of years fit somehow into the "balance of nature," and that aliens are more likely to spread and get out of hand and become serious pests.

Experience in other parts of the world gives some support to this point of view. In Australia and New Zealand plants like brambles and ragwort have become major pests, whereas in Britain they seldom if ever present such a serious problem. The usual explanation is that in Britain they are kept in control by the many insects which prey upon them; in the southern hemisphere they have been introduced without these insects. Further support is given to this view by the fact that some of these introduced weeds have been successfully controlled by introduced insects, and the whole subject of the biological control of weeds is being studied in a number of laboratories. So far the most strikingly successful cases of control have been of introduced weeds which invade pasture; annual weeds of arable crops seem less susceptible to these methods.

In Britain, though there have been many successful invasions by alien weeds, as has been so well described by Sir Edward Salisbury, none of these has presented a major economic problem. The important weeds in cereals or other arable crops are all natives; they were probably uncommon until agricultural developments produced conditions suitable for their multiplication. Sometimes it has been suggested that the actual strain of a weed species which is troublesome is an introduced one, but this is difficult to prove.

As mentioned above, plants introduced into new countries may become rampant weeds. The explanation about the lack of parasitic insects is perhaps not the whole truth, for from time to time plants which have existed in Britain for long periods of years as comparative rarities may suddenly become very common and in fact be reckoned as troublesome weeds. A familiar case is that of the rosebay willowherb, *Chamaenerion angustifolium*. This was a comparatively rare plant until some fifty years ago; to-day it is often the first to appear on a bare piece of ground in the country or even in the heart of London. This change in incidence is sometimes explained as an introduction of a new strain, sometimes it is said to be the result of mutation and the production of an "aggressive" form. Whatever the explanation, cases like these may serve as a warning that new weed problems may easily arise in Britain, and give rise to pressures to use more powerful herbicides.

In general I am of the opinion that to-day weedkillers are not a major danger to wild life in Britain. For most purposes we are using less toxic and less persistent substances. On the comparatively small areas where no plant growth is tolerated we use persistent substances, but these are apparently almost non-poisonous to animal life, and they seldom "leak" from the places to which they are applied. Wrongly or carelessly used, herbicides can obviously do a great deal of harm, but as a rule they make no more drastic changes to the environment than do normal mechanical farming processes – and some of the more recently discovered chemicals may actually have less effect, for instance, on the soil fauna. Nevertheless, we must keep all these organic herbicides under constant observation. A great deal of work on their toxicity has been done, usually with satisfactory results, and the obviously dangerous substances have usually been kept off the market. Nevertheless we cannot ever be 100 per cent certain that even the most seemingly safe herbicide has no dangerous properties, especially to man and animals which are in contact with it over periods extending to many years. It is well known that chemicals not dissimilar to herbicides are carcinogenic, and others have been shown to increase the rate of mutation in plants and animals in the laboratory. There is no proof that any herbicide in its normal use has ever caused cancer in man or in an animal, or that any harmful (or other) mutations have resulted, but it is right that these possible risks should always be borne in mind.

I mentioned at the beginning of this chapter that weed control is not always an economic proposition, for the cost of labour and chemicals may be greater than the value of the additional yield, and in fact there is sometimes no additional yield at all. There are various weeds which, in fact, do little harm. Low-growing plants like pansies and spurreys usually appear in a cereal crop when it is several inches high, and they do not compete seriously either for light or nutrients, so the yield may not be affected. They do not make harvesting difficult, as combine harvesters cut off the straw above the top of such plants. The aim of the conservationist is to ensure that herbicides are used only when really necessary, and he would urge the farmer to be prepared to accept some minor degree of inefficiency in the interests of wild life – and, perhaps, in the long run, of human health. Routine herbicide spraying should be discouraged; if sprays were applied scientifically, only when they were required to ensure an economic advantage, both farmers and conservation interests would gain.

CHAPTER SEVEN

FUNGICIDES

Fungi are plants without chlorophyll. They cannot therefore use light energy to convert carbon dioxide and water into energy-rich carbohydrates, but must depend on green plants to manufacture such substances. They live as saprophytes on decaying organic matter, or as parasites at the expense of living plants or of animals. Saprophytic fungi are important in helping to break down dead plant and animal material, for they eventually release most of the nutrients and so contribute to the fertility of the ground. Edible species such as mushrooms and truffles are of some economic value, and poisonous toadstools cause a few deaths in Britain every year. A saprophytic fungus like that causing dry rot in timber is often thought of as a parasite, as it damages man's property. The species which rot wood and discolour clothes or walls do a considerable amount of damage, and attempts are made to control them by various means including selective poisons or fungicides. An important industry has grown up for this purpose. Timber is impregnated to reduce its susceptibility to attack, and affected houses are treated. I know of no instances of wild life being harmed by this treatment, though wood-eating insects do not thrive on treated timber. Chemicals are only used extensively in old and badly designed buildings. With proper construction and good ventilation, saprophytic fungi do not present a serious problem.

Parasitic fungi are of the greatest economic importance. A few species attack living animals; in man they cause ringworm, and domestic animals and stock are similarly affected. Wild mammals of many orders are also attacked, though little seems known about the importance of such diseases. The fungus lives in and at the expense of the epidermis. Various fungicidal medicaments have been successfully developed. These are relatively non-toxic to other forms of life, and contamination of the environment has not been reported. The ringworm fungi are somewhat susceptible to sulphur. The "leprosy" of Naaman the Syrian (Kings II, ch. 5, v. 14) has been diagnosed on occasion as due either to the itch mite *Sarcoptes scabiei* or to the ringworm fungus; both might have been alleviated by the sulphurous springs which feed the Jordan. Atmospheric pollution with sulphur, which discourages rose mildew in industrial centres (see p. 36) could have a similar effect on fungi attacking the skin of animals, though this subject does not seem to have been extensively studied.

Fungi parasitic on plants are also of major economic import-ance. They cause serious diseases of cereal crops and fruit trees, and damage many plants in gardens, nurseries and glasshouses. Many of these diseases can be controlled, either by chemicals or by other means. Very large quantities of fungicidal chemicals are used in agriculture and horticulture every year, and these could sometimes have side effects harmful to wild life.

As a general rule fungi are much more easily harmed by copper than are green plants. High concentrations of copper salts do in fact damage all plants, but when carefully applied these sub-stances act as selective poisons, and there is often a reasonably wide margin of safety between the dose lethal to disease-causing fungi and that which is phytotoxic to the host crop. **Bordeaux mixture**, made by mixing copper sulphate, quick lime and water, has long been used to protect vines in Europe against fungus attack, and it, or a similar mixture, is still often widely applied to potatoes to combat blight, and to fruit trees to control apple and pear scab. Fungicides containing **sulphur** are also widely used as dusts and washes, and new organic fungicides are now being used in increasing amounts.

Fungicidal sprays and washes have seldom been proved seriously to endanger wild life. Copper and sulphur produce stable residues which remain more or less permanently in the soil. When these reach high concentrations, they have their effects, but the only case I know of which has been properly established and investigated concerns earthworms in orchards (see below). Rivers and ditches often become contaminated by the run off from orchards and potato fields. Algae and fish are killed as a result, but few serious incidents with deaths continuing over a long period have been reported. The effect of these toxic substances on many species of wild fungi must be considerable, and would repay study. The whole balance of plant life, and of the breakdown process in the soil, must be upset locally, and sometimes widespread effects are possible, even if they usually pass undetected. However, it seems probable that fungicidal sprays seldom play a part of major importance in damaging wild plants and animals, and there is no evidence that they are a major cause of environmental pollution. They have not been found to be transmitted from animal to animal in food chains, nor is there evidence of dangerous concentration by animals towards the end of the food chains (i.e. in predators such as sparrow hawks or stoats).

As mentioned above, the only well authenticated case of the effect of copper-containing fungicides on wild life was found in East Anglian orchards. One particular orchard had a long history of intensive spraying with copper compounds, and analyses showed that the surface litter contained as much as 0·2 per cent of the metal. This seemed to have no harmful effect on the grass, and the trees, mature Bramley Seedling apples for the most part, gave excellent crops. However, the soil was almost devoid of animal life. Earthworms, including the large *Lumbricus terrestris*, were not to be found. This species of worm feeds mainly by pulling leaves and other vegetable matter down into its burrows. In the orchard studied, the upper layer of the soil was peaty due to an accumulation of vegetable matter, and apple leaves remained on the surface throughout the winter. Neighbouring orchards, similar except that much less copper had been applied, had flourishing worm populations, and leaves and grass were rapidly removed from the surface by being drawn into the burrows. Although these mature apple trees did not apparently suffer from the presence of this large accumulation of copper, or from the absence of earthworms and other soil animals, it is likely that young growing trees would be affected. Also it is possible that in most orchards worms help to reduce the incidence of apple scab. This fungus over-winters on the fallen leaves, and the spores which bring back the disease to the trees develop on these leaves next spring. When worms bury and eat the leaves far fewer spores are liberated.

The soil fauna must be affected in many orchards, and in fields where potatoes are frequently grown and sprayed with copper compounds to control blight. I have no doubt that further research will reveal other cases, perhaps not so extreme as those in the orchard mentioned above, but covering wider areas. The effects will be long-lasting, as the copper rem,ains for years. The results may not be immediately apparent and may not be serious to arable agriculture, but the cyclical processes by which organic matter is incorporated in the soil, and by which dead plants are broken down, will clearly be disorganised. The soil fungi which we know play an important part in this cycle are rare in sprayed areas. Modern organic fungicides which are not so persistent are likely to have fewer effects of this kind, but we know too little about their other properties so they also need further study.

Cereal crops are susceptible to serious fungus diseases. Some of these are seed-borne. For instance the spores of the fungus causing bunt in wheat adhere to the grain. They germinate at the same time as the grain, infect the reproductive shoot and destroy every grain, leaving only a mass of black spores. Not only is the yield lowered, but the value of the remainder of the crop, which may be seriously contaminated and badly tainted, is greatly reduced. Bunt and other seed-borne fungi have been controlled by various means. Some are susceptible to heat, being killed by temperatures which do not harm the seed. **Formalin** and other chemicals have been used. To-day seed dressings containing mercury are preferred. At one time inorganic substances such as calomel were used, but to-day **organo-mercury** compounds are usual and many merchants automatically treat all their seed, so that untreated seed is difficult to obtain. Organo-mercury compounds are remarkably toxic to fungi, and although

in large doses are deadly to fish, birds and mammals, cases of
poisoning have seldom been reported in Britain.

A number of different organo-mercury compounds has been
used as seed dressings. Remarkably small amounts have proved
effective in controlling fungus disease. Thus as little as one ounce
of an organo-mercury preparation, containing perhaps only one
per cent of mercury expressed as the metal, may be applied to
one bushel of wheat, weighing approximately 63 lb. If this
wheat is sown at a normal rate of three bushels to the acre, only
about a milligram of mercury will be added to each square yard
of the field. If this is eventually incorporated into the top six
inches of soil, it will yield only about one part of mercury to
some hundred million parts of soil, far less mercury than occurs
naturally in most soils.

Farm stock, both cattle and poultry, have often been fed on
grain treated with organo-mercury fungicides, and few animals
have suffered; in fact it has been suggested that such treated
grain is advantageous to health and promotes more rapid growth.
However, there have been cases where poultry have suffered
when fed *exclusively* on dressed grain, though when this is only
part of the diet little harm is reported. Game birds, in captivity,
readily eat grain dressed with organo-mercury fungicides (though
insecticide dressings, particularly of BHC, make the grain unpalatable, see p. 133), and in most tests it has had no reduction
of longevity, egg-production or fertility, though forcible feeding
with large amounts has given some toxic effects. In Britain
many birds, including pigeons and pheasants, have been observed
digging up and eating dressed grain, and no poisoning from
fungicides has been observed, in sharp contrast to the effects of
seeds dressed with some insecticides (see p. 138). In Britain we
have, so far, no evidence that mercury is concentrated in food
chains.

However, recent work has shown that in Sweden there is
evidence to suggest that mercury poisoning is common among
wild life, and there is considerable anxiety that man may also
be affected. This poisoning is thought by some workers to come
from seed dressings. It is possible that the situation in Sweden

differs from that in Britain because different compounds are used. Much of the corn in Sweden was dressed with methyl-mercury dicyandiamide, while in Britain and most other countries phenylmercuric urea is more often used. Some Swedish seed merchants have also used rather high rates of dressing. Apparently these methyl organo-mercury compounds are the more toxic but this does not explain all the findings. Pheasants with over twenty parts per million of mercury, dead from mercury poisoning, have been found, and mercury seems to have accumulated in pike, hawks, owls and other predators. Specimens of all these have been found, dead, with all the symptoms of mercury poisoning. Few corpses with high residues of mercury have been found in Britain, but the metal has not always been estimated where other poisons were suspected. The position in Sweden is complicated by the widespread use of mercury in industrial processes; many scientists suggest that this is the major cause of contamination, though others incriminate seed dressings, as poisoning occurs in agricultural areas, though it has also been reported near industrial undertakings which use mercury. The Swedish findings are causing workers in other countries to look for similar results, in case seed dressings used there are more dangerous than has previously been realised. The effects of chronic pollution with very small amounts of mercury on animal tissues, and on soil fungi are also being investigated. The situation is complicated by the normal presence of mercury in all soils and tissues, often in amounts of a greater order of magnitude than would be expected from the minimal contamination of seed dressings. I myself do not think that organo-mercury seed dressings, as used in Britain to-day, are a serious danger to man or wild life.

There are some important fungus diseases which cannot as yet be controlled by chemicals. Take-all in wheat and barley, in which the fungus remains in the soil after cropping, and in spring attacks the roots of the young plants, causing the plants to die early and produce "white-heads" with almost no grain in the ear, is one. Eye-spot, in which the fungus attacks the base of the stem, and may cause the whole crop to fall over and so make harvesting impossible, is another. These diseases are usually only serious when cereals are grown without a rotation of other crops. Various cultural changes can keep down their incidence.

Most farmers would prefer to grow crops which are immune to disease. Considerable progress in developing resistant varieties has been made. Some varieties of wheat are less susceptible to eye-spot than others, though complete immunity has not yet been developed. Wart disease of potatoes, particularly important because the fungus remains alive in the soil for many years ready to infect the next potato crop, and because no satisfactory chemical treatment has been devised, attacks some varieties little if at all. Plant breeders are making considerable progress in developing resistance in several other crops.

It seems then that although fungicides are often very poisonous, most of them cause little obvious damage to wild life. Already less toxic and less persistent chemicals are being used, and the tendency is to reduce rather than to increase doses, though a larger area of the country is treated each year. Certain diseases have already been conquered by breeding resistant or immune varieties of crops. Some caution is, however, necessary. The effects of small doses of organo-mercury and other fungicides over long periods need to be observed, and it is possible that effects on saprophytic soil fungi are more important than has been realised. Residues of copper have accumulated in some soils, and may be more widespread than has been realised. Finally, though the situation is at present apparently satisfactory, and if anything changes are in the right direction, there is always the possibility of the appearance of new and virulent strains of disease-causing fungi, and of strains which are resistant to the chemicals in present use. If this happened, there might be pressure to use more dangerous substances in larger doses.

CHAPTER EIGHT

INSECTICIDES AND INSECT CONTROL

In the introductory chapter the subject of the control of insect pests was briefly introduced. Insects compete with man in various ways. They infest his person and his habitation, they eat his crops, and they carry and transmit micro-organisms and other parasites which cause diseases to man himself, his domestic animals and his crop plants. Man tries to control these harmful insects; he often succeeds, but sometimes this success has unfortunate side effects. It is important to remember that only a small minority of insects are recognised as harmful. The majority can be considered as "indifferent" in that they exist without apparently having any direct effect on man or his activities. A third group is beneficial, either because, like the honey bee or the lac insect, they produce substances of economic value, or because they are biologically useful. This biological value may be because they pollinate fruit trees and other crops, or because they help to control pest species. Entomologists have the difficult task of controlling insect pests without damaging valuable, or even indifferent, species. Incidentally, most of us now believe that many of the apparently "indifferent" insects may have some biological value, and that they should be protected unless they are known to be harmful. Few butterflies have a cash value, but our country would be much duller without them. Even "pest" species are not always entirely harmful – they may under some circumstances be actually beneficial in that, by eating plant materials and breaking them down, to be returned to the soil, they form an important part in the essential cycle of nutrient substances. More often than not they exist in small numbers without doing appreciable economic damage, and unnecessary control is a waste of money as well as a source of environmental pollution.

The most direct method of insect control is by the use of insecticides. These are substances which are poisonous to insects. They are also poisonous to other forms of life, but they may either be much more lethal to insects than to mammals (including man) or it may be possible to apply them in such a way that insects receive lethal doses while other forms of life escape. Different species of insect are usually different in their susceptibility to an insecticide, and some success has been achieved in producing substances poisonous to pests and less deadly to beneficial species. However, almost all insecticides will kill other animals and plants if they are used improperly, and most can unfortunately have at least some harmful effects even if used with all possible precautions.

Many different substances and methods have been used to kill insects. Common household products like vinegar and paraffin oil (kerosene) are applied to the human scalp to control head lice, and are poured into cracks in bedsteads and other furniture to kill bedbugs. They are not very effective in either case, but neither are they totally ineffective. High concentrations of substances which are usually considered innocuous can act as poisons, and when insects are subjected to these treatments they receive these large amounts under circumstances where escape is difficult. It is unlikely that other forms of life are often accidentally affected, though kerosene and other oils are still used to control pests on fruit trees and to "oil" water to kill mosquito larvae. In this latter case mosquitoes are usually controlled by small amounts of oil which seldom harm fish or birds using the same water, and few other forms of insect life appear to be affected. Soaps are also among the many substances with some insecticidal properties. Aphids on roses have long been discouraged to some extent by pouring the washing-up water over the bushes! Various soap preparations have been used against a variety of insects, but in recent years this practice has declined. The fact that these techniques seldom damage wild life may encourage their introduction in due course.

Heat, obtained by the application of steam, hot air or hot water, is lethal to all insects, provided that a high enough temperature is obtained in the exact spot where the insect is lurking. The actual temperature which will kill most insects, and even their eggs (which are sometimes more resistant to heat, as they are to poisons also), is much lower than many people imagine; 50°C. (122°F.) for a few minutes is effective in most cases. As mentioned above, these temperatures must reach the place where the insects are, and it is surprising how long it may take to get up to 50°C. in the middle of a pile of bedding even when the air in the room is much hotter. Heat is still used to kill lice in clothes and bedding, and to eliminate insects and mites from grain and other stored foods. In countries with a hot summer and a cold winter bedbugs have been eliminated from empty houses by turning on the central heating in warm weather. Granaries have been designed in the tropics where the temperature is deliberately allowed to rise and kill insects infesting the contents; care must be taken to prevent too high a temperature, harmful to the grain, from being reached. Seeds have often been freed from fungi, and bulbs and some other plants from nematode parasites and even pathogenic viruses, by exposure to carefully controlled temperatures. Heat is clearly a safe "insecticide," having no unwanted effects outside the area under treatment. Unfortunately it cannot usually be used against infestations of growing plants, which would usually be killed as readily as the pests.

Many of the poisons familiar to toxicologists have been used at some time or another to kill insects and other pests, and some have proved quite successful for this purpose, though the danger to operatives and others has been great. **Cyanide,** usually in the form of hydrocyanic acid gas, has served as a fumigant in buildings against bedbugs and wood-boring insects. It was extensively used from 1886 onwards in California against the cottony cushion scale insect (*Icerya purchasi*); canvas tents were placed over the infested citrus trees, and a high concentration of HCN was liberated inside. The method was at first highly successful, though after a time failures were reported. These were found to be due to the selection of strains of the pest which were unusually resistant to cyanide; this was the first case of resistance to an insecticide, one of the great dangers to pest control, which was ever reported. Cyanide is phytotoxic, but usually a much higher concentration is needed to harm trees than to kill insects. However, scale insects became so resistant that they survived when the trees were killed by severe exposure to cyanide. Cyanide is so obviously dangerous, and such a well recognised poison, that it is seldom difficult to enforce safety precautions. Nevertheless, numerous human deaths have occurred. Wild life has rarely been affected. Beneficial and other insects on treated trees were killed, but there was little effect outside the orchards. Birds which fed on insects killed by HCN were seldom harmed. Cyanide is little used as an insecticide to-day, except as a fumigant in glasshouses, and then with such care that it can seldom be a danger. The poison is rapidly broken down, so there is no trouble from persistence. The one common use of cyanide to-day is to gas rabbits in their burrows (p. 170). Badgers and foxes are sometimes killed, perhaps by accident when their earths are adjacent to those occupied by rabbits, and sometimes deliberately. Insects, worms and other invertebrates in the soil will also be killed over a limited area, but the side effects of killing rabbits by HCN are probably small. Rabbits poisoned by this means can certainly serve as food for maggots a few days after they die, and normal flies emerge from the corpses.

Arsenic is another general poison which has been widely used against insects (and against weeds, see p. 78). As early as 1840 various arsenical compounds were used, and in 1867 "Paris Green" (containing copper arsenite) was successfully used to control the Colorado potato beetle in the Eastern United States. Numerous different arsenical compounds were used, but most were abandoned because of their toxicity to man and because they were also phytotoxic (a disadvantage when used to control insect pests on plants). The only substance in this group which is widely used to-day is lead arsenate, which is still listed in the Agricultural Chemicals Approval Scheme, for use against various caterpillars on fruit trees. It has also been used

against soil pests such as leather-jackets and wireworms, mainly
on restricted areas such as golf greens where costs are not as
important as on an economic crop. Surface-casting earthworms
have also been eliminated from greens by its use. Many people
think that the use of arsenic in agriculture and horticulture has
been banned; as indicated on p. 78, it is sodium arsenite, used as
a herbicide, that is no longer available. This substance was
undoubtedly much more dangerous to operators, and possibly
to wild life, than the much less soluble lead arsenate. Nevertheless, lead arsenate is a very poisonous chemical, and must not be
applied for some six weeks before a crop is marketed. The Food
and Drugs Act, 1955, also lays down that it is illegal to sell any
food containing more than one part per million of arsenic, or
more than two parts per million of lead. At this level, at least a
ton of apples would have to be eaten over a short period to give
the eater a lethal dose (though the chronic effect of smaller
amounts accumulated over a long period might have some harmful effects). The content of lead arsenate in soil after worms have
been killed remains at a lethal level for many months, and even
for years in some situations. There is some leaching of the poison,
and run off water can kill fish and other forms of life. Birds which
eat many poisoned worms are likely to be affected, though
records of this are not at all common. I think that knowledge of
the properties of arsenic is so widespread that here, as with
cyanide, care in application has prevented a great deal of harm
from being done.

Fluorine, already discussed as a dangerous poison resulting
from industrial pollution (see p. 39), also has insecticidal properties. Before the introduction of the chlorinated hydrocarbons
(DDT, etc., see below), it was used against various domestic pests,
including cockroaches and Pharaoh's ants. Sodium fluoride was
mixed with flour or sugar and used as a bait. It acted as a
stomach poison, and was also to some extent absorbed through
the cuticle. Attempts were made to use various fluorine compounds against insect pests of crops, but they were never widely
adopted. The risk to wild life of fluorine used as a pesticide is
very slight, particularly as it has now gone almost entirely out

of use. Were it reintroduced the effects would be similar to those from industrial pollution, but probably over even smaller areas.

Poisons extracted from some plants are insecticidal. The first such substance to be widely used was **nicotine,** extracted from tobacco. Water in which tobacco leaves had been soaked was used as early as the middle of the eighteenth century against pests on fruit trees. Until recently nicotine was the only efficient poison against greenfly and other species of aphid. Nicotine acts as a fumigant, and also as a contact poison which penetrates the insect's cuticle. Its efficiency as a fumigant is a measure of its volatility; plants effectively treated quickly lose any poisonous deposit and become safe to eat, so crops can be marketed a few days after spraying. Nicotine can be used to fumigate glasshouses which can be used, even by entomologists wishing to rear insects in them, after about forty-eight hours. It is a substance almost completely devoid of any phytotoxicity, so that however used no damage to vegetation takes place. It is very poisonous to man and to mammals; carelessness has caused serious accidents to users, and farm stock has died from nicotine poisoning. It is, however, a substance which is unlikely ever to contaminate the environment, or seriously to endanger wild life.

Other insecticides of vegetable origin mostly differ from nicotine in being far less toxic to man and other mammals than they are to insects. The two most widely used are rotenone and pyrethrum. **Rotenone** is the active substance of the root of the tropical plant *Derris elliptica*, and the ground root is itself used and is generally simply spoken of as "derris." In fact, quite a number of other leguminous plants also contain rotenone and are used in the same way as derris. These substances were originally used as fish poisons in Malaya, South America and elsewhere, and their wide use as insecticides is recent, though limited use of derris was made over a hundred years ago. Rotenone breaks down fairly quickly and so is not a persistent insecticide. It is remarkably non-poisonous to mammals and birds, though it is deadly to fish. During the 1939-45 war it was the active ingredient of the highly-effective preparation "A.L.63" which was used against lice in Italy and in fact achieved a considerable degree of control of the typhus epidemic in Naples before DDT was introduced and finished off the outbreak. Although many human beings had their undergarments impregnated with rotenone little harm was done. In a few cases the skin became sensitised to the derris and some dermatitis occurred, but this happens with many plant products which are not poisonous to man (e.g. pollen causing "hay fever" also causes dermatitis in sensitive individuals) and it may not have been the rotenone itself but some other substance in the plant which caused the trouble. Rotenone is still used in horticulture, where its non-persistence and lack of toxicity is useful. The dust is blown on to plants (e.g. raspberries against raspberry beetle) and may kill a few other insects but widespread damage does not occur and wild life is hardly at risk. Ornamental fish in ponds in gardens may occasionally suffer but I have never heard of a case of a stream or river being seriously polluted by derris.

Pyrethrum is probably the most widely used insecticide of vegetable origin. It is usually prepared from the flowers of *Chrysanthemum cinerariaefolium*, which has become an important crop in Kenya and several other tropical and subtropical countries. A number of other related plants can also be used to prepare similar substances. Pyrethrum is said to have been used on a limited scale by Caucasian peasants for centuries, and has been an article of commerce for over a hundred years. The most striking feature of this insecticide is that it gives a remarkably rapid "knock-down," that is, when an insect is treated it is quickly paralysed. After a small dose, the insect may recover, and for this reason some preparations are marketed with pyrethrum to give an immediate effect, and some longer-lasting substance (DDT for instance) to finish off the victims. However, pyrethrum alone is a most effective insect poison; paralysis of an insect is accomplished by a very low dose indeed, and death is caused by a slightly higher, but still, in comparison with other insecticides, a low dose.

It is unfortunately rather expensive to produce pyrethrum, so it has been used more often in public health work and to control disease-carrying insects like malarial mosquitoes than in agriculture. Sunlight and air cause rapid breakdown and loss of toxicity, though pyrethrum is more stable in oil, and a solution painted on the walls has been used to control warehouse pests. During the 1939-45 war supplies of pyrethrum ran very low, just when it was needed for mosquito control on tropical fronts, and this was probably responsible for stimulating the interest in synthetic chemicals which culminated in the realisation of the properties of DDT.

Pyrethrum appears to be entirely devoid of any phytotoxic properties, so it will harm neither crops nor plants which are accidentally treated. It can in large doses be poisonous to mammals, but in practice accidents seldom, if ever, occur. A man would have to consume about an ounce of the active ingredients to suffer severe harm, and it would be possible, though hardly pleasant, to gorge on the flowers without noticing the effect. It does not seem to be any more toxic to birds. Its effect on vertebrate wild life is therefore minimal. Any insect which comes into contact with pyrethrum is likely to be killed, so some beneficial or aesthetically desirable species will often perish, but the lack of persistence means that long-term damage will seldom occur. From the point of view of wild life conservation, pyrethrum is very nearly the ideal insecticide. Incidentally attempts are being made to synthesise substances chemically similar to pyrethrins, in the hope of obtaining large supplies of a desirable insecticide cheaply. Some people seem to feel that "synthetic pyrethrins," being allied to a "natural" insecticide, will therefore be more acceptable than substances like DDT. Unfortunately some of these synthetic pyrethrins have been rather poisonous and potentially as dangerous as other synthetics with no affinity with pyrethrum.

DNOC as a herbicide has already been discussed (p. 79). This very poisonous substance was, however, first used as an insecticide. As long ago as 1892 it was used to control the Nun moth, a pest of forest trees. As DNOC is sufficiently phytotoxic to be used as a herbicide, its insecticidal value is obviously limited. It has nevertheless had wide use as a "winter wash," sprayed on to fruit trees in winter (when there is no foliage to damage) to control aphid and mite pests. It was found to be effective against the resistant eggs deposited by some pest species, and for this reason is still used. Wild life in an orchard is at risk when DNOC is sprayed, but few birds seem usually to suffer as they are scared from the bare branches by all the tumult which accompanies application. Beneficial insects are killed, and the elimination of predators by DNOC may have contributed to the appearance of the red spider mite as a pest (see also p. 129). Drift into surrounding fields and woods may do damage, but as DNOC can only be used as an insecticide in winter, because of its phytotoxicity to the trees or plants infected, this damage is not serious. This substance has probably done most harm as a result of application as a herbicide, where it has been applied over wide areas in spring when other plants and animals are more susceptible.

This section summarises insecticide use up to the 1939-45 war, though some of the new substances dealt with in the later sections had been introduced on a limited scale, particularly to tackle medical rather than agricultural problems, before 1945. Some of the early pesticides, such as arsenic and cyanide, are deadly poisons, and potentially offer serious dangers to wild life, but their very toxicity has probably prevented their abuse in most cases, and they are now generally well controlled. The insecticides of vegetable origin like derris and pyrethrum have many advantages, and are still used. They are almost completely devoid of phytotoxicity, and are not very dangerous to mammals and birds. They are likely to be used even more widely if they can be produced cheaply, or if they are not replaced by synthetic substances with similar properties.

THE ORGANO-PHOSPHORUS INSECTICIDES

The insecticides which have been most widely used since 1945 fall into two groups, the chlorinated hydrocarbons, often called the organo-chlorine insecticides, of which DDT is the best known, and the organo-phosphorus compounds. I am dealing with the latter group first, because their wide introduction into agricultural practice caused considerable alarm at a time when DDT and similar substances were considered to be comparatively "safe." Actually the insecticidal use of DDT preceded that of Parathion,[1] the first organo-phosphorus insecticide to be used, but it was mainly as a medical insecticide whereas Parathion was from the beginning used in agriculture.

Parathion and the other early organo-phosphorus insecticides are efficient killers of pest insects, but they are also very poisonous to mammals, including man, and to birds. Until their toxicity had been properly realised, and suitable precautions, which included wearing protective clothing which entirely covered the bodies of the operators, had been devised, a substantial number of human deaths occurred. Their use in tropical conditions has been particularly hazardous, first because they have been used by untrained staff who have not understood the danger, and secondly because in hot conditions protective clothing is, at best, extremely uncomfortable, and at worst may cause death from heat stroke. Some operators have suffered the worst of both worlds. Protective clothing, not properly buttoned up because of the heat, but still sufficient to cause distress, has allowed skin contamination with the poison, which has proved exceptionally dangerous under these conditions.

The mode of action of organo-phosphorus insecticides has been studied by many workers. These substances appear to inhibit the working of the enzyme cholinesterase, and so act on the nervous system. However, their action on different animals varies, and the original comparatively simple picture of their action and toxicity clearly does not tell the whole story. As with most poisons, we do not fully understand how they work.

---

[1] There is some doubt as to which insecticide was used first. The *Insecticides and Fungicides Handbook* says that Parathion was the first organo-phosphorus substance to find insecticidal use, being introduced in 1944. TEPP, another organo-phosphorus compound, is said to have been used as a nicotine substitute in 1939. However, Parathion was the first organo-phosphorus insecticide to be widely used in British agriculture.

Toxicity to rats of some organo-phosphorus insecticides

	LD_{50} mg/Kilo
TEPP	1
Phorate	3·7
Parathion	6
Disulfoton	12·5
Phosphamidon	16·8
Dichlorvos	80
Demeton-methyl ("Metasystox")	40–180
Menazon	1950
Malathion	480–5,800

When the toxicity of such organo-phosphorus compounds as Parathion and **TEPP** (which is probably the most poisonous substance ever used on farms, for one ounce could poison fatally nearly 500 men) was realised, their danger not only to man but also to wild life caused much public concern. As already mentioned, protective clothing is now prescribed for operators using these substances. The danger to wild life has fortunately been much less than was originally feared. Beneficial insects in a crop which is sprayed are usually as susceptible as the pest species which are the targets of the operation. Any birds or mammals wet by the spray have little chance of survival. However, these compounds quickly break down and cease to be dangerous, and birds entering a sprayed area a few hours after application usually survive. Careful study has not shown that the very toxic organo-phosphorus insecticides have ever been responsible for any major catastrophe affecting British wild life. Some birds and mammals undoubtedly died, bees were on a number of occasions wiped out in orchards which were sprayed at the time when the workers were visiting the blossom, but even this was found often to be avoidable if the spray was carefully applied at a time when the bees were not visiting the trees, for instance in the evening. The comparative harmlessness of most of these very poisonous substances is due to their instability; they break down rapidly, so that even a day after application a crop is comparatively safe.

Nevertheless, we were probably lucky that Parathion did so little harm to British wild life, as it was much more destructive in America, particularly in California. There it was used against mosquito larvae, and large areas of rough ground and marshes were sprayed. Many game birds perished. Incidentally it proved difficult for some time to diagnose the exact cause of death, because no residues of these comparatively unstable poisons were detected in the corpses (a very different situation from that found with the chlorinated hydrocarbons; see p. 148). Modern analytical techniques will now detect residues, even though these are not themselves harmful.

Parathion acts as a contact poison, killing insects (and other animals) whose skins are wetted with it. It is also moderately volatile, and the toxic gas in a crop kills other insects. At first this effect was thought to show a so-called systemic effect, that is that the poison was translocated by the crop plant, killing insects which fed upon them even on parts of the plant which had not themselves been touched by the spray. In fact Parathion is not a systemic insecticide, though many of the newer organo-phosphorus compounds are, and much research has been directed to their discovery.

Systemic insecticides have two advantages. As already mentioned, the whole surface of a plant can become lethal to insects feeding on it, even if only a part of the plant is treated, so there is no need for a complete cover when a spray is applied; this means that the amount used can be reduced, with economy to the farmer and less danger of contamination of the environment. A systemic insecticide may even be applied to the roots with minimum risk to predators. Secondly, it is possible for only the pests, actually feeding on the crop, to be poisoned, while other insects sheltering among the leaves may be unaffected. This means that beneficial insects, for instance predatory ladybird beetles, may survive, and may even be able to finish off those aphids which have escaped the insecticide. This works if plant-feeding insects do not hand on a lethal dose of poison to the preda tors which eat them. Some organo-phosphorus substances break down quickly enough to obviate this danger.

Research has also been directed towards finding compounds which, though still as lethal to insects, are much less toxic to man and to warm-blooded vertebrates. This has proved surprisingly successful. **Malathion,** introduced in 1950, and **Menazon,** used first some ten years later, are both only about one two-thousandth part as poisonous as the most toxic organo-phosphorus compounds (e.g. TEPP) when tested against warm-blooded vertebrates. At first it was thought that such relatively safe chemicals were only negligibly dangerous, but unfortunately Malathion at least has caused human fatalities. This has been found to be due to the victims having previously been exposed to low, and apparently harmless, doses of Parathion (or to one of the other highly toxic organo-phosphorus compounds). The explanation seems to be that the first exposure damages the mechanism by which Malathion is rendered harmless within the mammalian body; Malathion is in fact highly poisonous to man, but it is easily and rapidly broken down to relatively harmless substances in normal individuals. This complex and cumulative effect of a relatively innocuous substance following a sub-lethal exposure to another chemical is an unlucky chance, but one that might not be uncommon to farm workers. The risk to wild life is probably much less. Short-lived animals, for instance beneficial insects, are unlikely to have normal life spans sufficient to allow them to be exposed to two different chemicals. Mammals and birds live much longer, and here the risk is a real one, though I do not know that it has been proved to occur in Britain. I must again stress that in Britain organo-phosphorus insecticides have seldom been used except on arable crops or in glasshouses, so the exposure of vertebrate wild life has been minimal. In the United States the much wider use of these insecticides has meant a far greater risk to wild life, and cases of the cumulative effects of two organo-phosphorus substances have occasionally been reported. Nevertheless under most conditions – and fortunately most parts of the world have never yet been sprayed with any chemicals – Malathion is a safe and useful insecticide.

It is widely used in the tropics in malaria control, and for this purpose it is almost the ideal substance. It is used by amateur gardeners to control pests on fruit, vegetables and flowers; if this is necessary (see p. 201) then this is probably one of the safest chemicals to use.

Mention has been made of the search for **systemic** organophosphorus insecticides. In the late nineteen forties "Shradan" or OMPH (**octamethylpyrophosphoramide**) was introduced, and now there are many others, perhaps the most widely used being **demeton methyl** ("Metasystox"). These substances are all fairly poisonous to man and to warm-blooded vertebrates, though nothing like as dangerous as Parathion and TEPP. Farm workers are expected to wear protective clothing when handling these compounds in a concentrated state, or when using them inside glasshouses or other buildings, but there seems little risk to operators when spraying outdoor crops and here precautions are seldom taken. One of the most important uses of these substances has been to protect sugar beet from colonisation by the aphids which carry the virus causing the disease known as sugar beet yellows, a condition which if widespread, particularly if it appears in early summer, causes a catastrophic fall in the yield of sugar. In Britain at least it would seem that these systemic organophosphorus insecticides have done little harm to wild life. The main reason for this is that the chemicals are almost always sprayed where little wild life is present. Sugar beet is a comparatively recently introduced crop in Britain, and it may be for that reason it seldom harbours many birds or mammals. Some beneficial insects are no doubt killed, but even here harmful results are minimal. In a crop of this kind predators like ladybirds only enter in any numbers after their prey, the virus-carrying aphids, is established, and early spraying will thus avoid the predators. Also these systemic insecticides kill the aphids mainly when they suck the plant's poisonous sap, and they are often applied in rather dilute solutions; these may not harm all insects which are only subject to them at the time of application. There is some evidence that even if aphids which have taken sufficient insecticidal sap to make death inevitable are eaten by predators, these predators may survive, though some are undoubtedly killed by eating poisoned prey. Bees feeding on nectar have obtained fatal doses of systemic insecticides, so spraying should be timed to obviate this danger.

So far in Britain we have only used organo-phosphorus insecticides to any great extent on agricultural crops. In Canada one substance in this group, **Phosphamidon,** has been widely used to control the spruce budworm which is a serious pest of coniferous forest trees. Although Phosphamidon is poisonous, being about a third as toxic as Parathion, it was hoped that it would not be harmful to wild life, as preliminary test had shown that at the rate applied (half a pound per acre) fish in forest streams were unaffected. However, spraying the forest resulted in a serious mortality among birds, particularly warblers, which appeared to pick up the poison from the foliage. When the rate of application of the insecticide was reduced to a quarter of a pound per acre, it appeared to be harmless to the birds. This is just one instance of the way in which the exact dose of a toxic substance must be adjusted to control the pest without harming other organisms. Unfortunately under normal conditions it is difficult to obtain such results. Even with the greatest care and the best apparatus it is usually impossible to apply a spray absolutely evenly over a wide area. This means that some patches will receive little spray, and the pests will survive, while others will be overtreated so that many birds and animals are locally affected. The situation is aggravated by the tendency of many operators to try to ensure satisfactory pest control in even the under-sprayed areas by stepping up the dose applied to the whole area.

To control soil pests like the carrot fly, some organo-phosphorus insecticides (e.g. **Disulfoton** or **phorate**) have been used in granular form, the granules being sown adjacent to the seed. The intention is to produce the minimum soil pollution, and the maximum concentration of pesticide in the right place. The method has proved successful with little danger to birds, or to most of the soil fauna, for though worms and predatory insects near the granules may die, far fewer are killed than by spreading such a very toxic insecticide broadcast.   Unfortunately this method has sometimes harmed the operators, who breathe in dust from the machinery which places the granule in the soil. This danger can be avoided by the use of masks.

The question of insecticidal stability is discussed in some detail later in this book (p. 135). For the moment it is sufficient to say that this was originally welcomed, in the belief that a single application of a substance might render a crop "immune" to attack, and a building could similarly be kept free from rein-fection for a long period.   Many entomologists tried to make pyrethrum, otherwise the ideal insecticide, more persistent. However, when the first highly poisonous organo-phosphorus insecticides were introduced, their instability was clearly an advantage, as otherwise a treated crop would never have been fit for human consumption.   When less poisonous organo-phosphorus substances were produced, the need for instability was less important, and now chemists are searching for chemicals with minimal vertebrate toxicity but which do not break down quite so rapidly.   They are also trying to find chemicals which are highly specific in their toxicity, so that, in an ideal case, they only kill the pest and leave unharmed the predators which are helping to control it.

Dichlorvos, a volatile and unstable insecticide, is now being widely used to kill domestic insects (flies, mosquitoes, cockroaches) and those infesting stored products. Strips of plastic covered with the chemical are hung in confined spaces, and the insects are killed. The evidence available shows no harm to man or domestic animals and birds, and no poisonous residues have been detected in dangerous amounts on exposed food. The method cannot be used out of doors, so risks to wild life, including non-pest species of insects, are negligible.

One field in which organo-phosphorus insecticides may play a part is in protecting sheep from "strike," that is from the attack of blowflies: they lay their eggs in the fleece and the maggots which emerge then parasitise the flesh. This has been controlled by many sheep dips and sprays. A few years ago it was thought that the chlorinated hydrocarbon dieldrin, one application of which gave protection throughout the season, as compared with the need for repeated treatments with other chemicals, was the answer. Unfortunately (see p. 142) the dangers from dieldrin to man and to wild life have stopped its use for this purpose, and now organo-phosphorus compounds which may give protection for the season of exposure, that is several months, but which are not unduly stable, are being tried with some success.

It thus seems likely that members of this group of substances, which were originally feared by naturalists because they were so poisonous, may be produced which will combine effectiveness with reasonable safety, though the growth of resistance in pests, and many other factors, makes it unlikely that any one substance will be effective for more than a few years. No insecticide by itself is ever likely to give a complete and permanent answer to any pest problem. Finally, the danger that one organo-phosphorus chemical may "potentiate" another must never be forgotten. We know that a small dose of the apparently harmless Malathion may have a drastic effect on a man who appears normal but who has recently used Parathion. The damage, probably to the victim's cholinesterases, even from tiny doses of a single sub-stance absorbed repeatedly, may be cumulative, somewhat reminiscent of the damage caused by radiation, though of course the mechanism is quite different. Operators regularly using even apparently harmless organo-phosphorus insecticides should there-fore have regular medical examinations, for there is some evidence of the development of mental as well as other symptoms from chronic exposure.

THE CHLORINATED HYDROCARBONS
a: DDT

DDT was first synthesised in the laboratory in 1874, but at that time no one thought it had any economic value. Its insecticidal properties were not discovered until 1939, and it was only during the 1939-45 war that it was used to kill insects. I well remember when I first heard what was then a closely-guarded secret about this new "wonder chemical." During the 1914-18 war insect-carried diseases were a menace to all troops of all nations concerned. Lice, particularly during trench warfare on the Western front, and under conditions where hygiene was impossible in Eastern Europe, were found on almost every soldier, and the diseases (typhus, trench fever, relapsing fever) which they carried caused millions of deaths. In Southern Europe and Africa malaria was a greater danger than any human enemy, and deaths from disease far outnumbered battle casualties. It was not until the war had gone on for several years that serious attempts were made to control insects of medical importance. In 1939, however, we in Britain were more alert to the dangers of insect-borne diseases, particularly under crowded conditions, such as might be expected in civilian air-raid shelters as well as in trenches, and in Africa and the Far East. We even had a high-powered committee of leading scientists called the "Insecticide Panel" which sponsored research on these problems, and entomologists were organised even before the war began. However, we had few efficient insecticides at our disposal. Pyrethrum was the best substance available, but only in limited amounts and production could not be rapidly increased. Derris preparations were also valuable, but again only limited quantities could be produced. Every attempt was made to get the best results from these limited stocks. As mentioned above (p. 109), a preparation based on derris, A.L.63, was developed as a weapon against lice, but there was not enough for both the troops and the civilian population of liberated Europe. Pyrethrum was kept mainly against mosquitoes, but stocks were quite inadequate if insects had to be controlled on many fronts. Then DDT appeared. It seemed miraculous, for it killed insects at dilutions which, at that time, seemed greater than could be easily explained, yet it seemed practically harmless to man. Volunteers went for days in underclothes which had been impregnated with DDT; lice on their persons died at once, and, more important still, it proved impossible to reinfest them for weeks, though no more DDT was used. In many experiments of this kind no single volunteer showed any symptoms of poisoning. DDT was used in many ways, with similar results. Swamps were freed from mosquitoes by applying as little as one pound of DDT to the acre; men drank the water which had been so treated and were unharmed. Experiments showed that it *was* possible to poison warmblooded animals and birds with DDT, and that it was most dangerous if given in solution in oil, but in general the impression was that we had at last discovered the perfect insecticide, one which was quite safe to use if reasonable precautions were observed. Under war-time conditions, it really was very nearly perfect.

It may be significant to note that the acute toxicity of DDT is almost exactly the same as that of the drug aspirin. This does *not* mean that it is non-poisonous, for aspirin, taken 100 tablets at a time, is a common suicide drug. The same quantity (about an ounce) of DDT would be lethal to most men. There is, however, one very important difference. Many people can take smaller doses of ten or twenty grains of aspirin a day for a long period without any serious result. This drug is not retained in the body. But a substantial part of the DDT is stored in the tissues, particularly the fat, so a cumulative effect, not obtained with aspirin, can be obtained. Fortunately much of the DDT retained in the body is changed into the much-less-poisonous DDE. A substantial fraction is also excreted, and the larger the dose the higher this fraction. Nevertheless the effects of chronic exposure can be serious.

It is almost impossible to over-estimate the war-time importance of DDT. In the west, where DDT was available, louse-borne typhus was controlled, a thing that never happened under similar conditions in World War I. In areas occupied by the Germans, who did not use DDT for louse control, the disease was serious. Mosquito control was equally efficient. Results with malaria were not as dramatic for prophylactic atabrin had already reduced its incidence from about three attacks per man per year in New Guinea to an almost negligible figure, but other mosquito-borne diseases, as well as nuisances from house flies and other insects, were effectively contained. Most important of all, the knowledge that DDT was available allowed military operations to be planned in areas in the tropics where previously insect-borne disease would have made such operations unacceptably hazardous. DDT certainly helped the western powers to win the war.

After the war, entomologists continued in this optimistic frame of mind. We thought that we now had the answer to insect-borne disease, and in fact campaigns to eradicate malaria using DDT as the main weapon were launched, with every hope of success, in many countries. This substance undoubtedly saved many millions of civilian lives. At first DDT was not available in sufficient amounts for widespread use in agriculture, but production increased, and it was used with success on many crops. Increased food production in areas where famine is endemic once again alleviated what would otherwise have been great suffering. Control seemed so easy that some professional entomologists feared unemployment! When one reads the attacks which are being made to-day on the use of persistent organo-chlorine insecticides, it is difficult to recall the optimism of the late nineteen-forties, and the tremendous benefits these substances have conferred, and are still conferring.

In fact up till about 1954 few complaints were heard against the organo-chlorine insecticides, though some medical authorities were worried by residues, particularly when these were secreted in the milk of cows fed on foodstuffs which had been treated with insecticides. Fish poisoning in streams treated in anti-malaria campaigns was accepted as unavoidable. As long as DDT was mainly used to control human disease, little notice was taken even when such incidents occurred. In fact in Britain there is little evidence that even to-day DDT has caused any great amount of damage to wild life, and it appears to have been almost entirely harmless to man. This is not because we have always been wiser than others, but it is because we have not had comparable problems of insect damage to deal with. It is also possible that our authorities have not had sufficient money to carry out the extensive campaigns which have been a feature of insecticide use in the United States. Nevertheless, as will be shown later, although we may have avoided gross poisoning, our environment has been contaminated with amounts of DDT which cannot be accepted without question, and we may have escaped serious danger more by luck than by good management. And *some* organo-chlorine insecticides have in fact done considerable damage in Britain.

DDT, used with care, and at minimum doses, has successfully controlled insects in many countries, and has usually done so without any recognisable harmful side effects. Used in ways which we now consider unwise it has done a great deal of damage to wild life. The classic case which has often been described concerns the attempts, in America, to control Dutch elm disease. This is a condition caused by a fungus and carried by bark beetles. The disease was introduced into the United States more than thirty years ago, and has caused serious damage to elm trees, which are important shade trees in many suburban areas. In some districts the elm trees have been almost wiped out, so there is no doubt of the seriousness of the problem. The beetle has proved very difficult to control, but if it could be exterminated the disease would be at least on the way to being controlled. From 1955 onwards elm trees have been sprayed with DDT in several different areas. High doses of the chemical, as much as 25 lb. to the acre, were used. In almost all cases birds, particularly the American Robin, were killed, and there are several authentic records of whole bird populations being wiped out. It seemed that the DDT got on to the soil, where it poisoned earthworms, which in turn poisoned the birds. Poisoned caterpillars also contributed to the situation. Direct poisoning of the birds themselves occurred but did not seem to be the major problem.

Many other incidents similar to that just related have been recorded, in the United States and in other countries where DDT has been used over wide areas of territory. In Britain the chemical has seldom been used in this way. One of the few British pests which might be controlled by similar means is the pine looper caterpillar, which is occasionally a serious pest of forest trees. In most years no serious damage occurs, so spraying should only occasionally be necessary. Aerial spraying with DDT against the pine looper has been done in the pine forests in the Cannock Chase area. In 1963 over a thousand acres were sprayed from the air in late summer with 1 lb. of DDT to the acre, and the results have been carefully studied. The caterpillars were successfully controlled. No bird casualties were reported; the operation was timed to avoid the nesting season, when results would probably have been different. Birds of several species, including tits, were sampled before spraying, and were found to have low DDT residues in their tissues. Soon after spraying further sampling showed a rise in these pesticide levels, but not to such a height as to cause deaths. A year later birds still contained DDT and its metabolites, at levels slightly higher than before the spray was applied. It is of course possible that some birds died and were not found, but observations suggest that many simply moved away at the time of spraying. In later years breeding success does not seem to have been reduced. One interesting point concerns insecticide residues in the soil. This area had been sprayed several years before 1963; quite substantial amounts of DDT remained in the soil, and this suggests that repeated spraying each year might have serious effects. Work on this experiment has not yet been completed, and all the results are not yet published, but it seems probable that it *is* possible to treat considerable areas of woodland with DDT without necessarily causing havoc to wild life, if suitable precautions are taken, minimum doses are used, and the operation is not too often repeated.

The greatest danger from the use of DDT over wide areas arises if water is contaminated. Damage to many forms of life may also occur when aquatic pests are controlled by chlorinated hydrocarbons. This is due to several causes. First, fish are particularly susceptible to DDT poisoning, so that fish deaths due to direct poisoning may occur immediately or soon after the insecticide has been applied. They may extract DDT, present in a low concentration, from the immense amount of water which is passed through the gills for purposes of respiration (see p. 49). Secondly, insects and other invertebrates, which are the main food of some fish, may be exterminated, so the fish are starved. Thirdly, the invertebrates may take up amounts of DDT which are not immediately lethal; when these poisoned animals are eaten by fish, the DDT may be retained in the bodies of the fish, which over a period obtain a toxic dose. To find clear-cut cases of the possible effects of chlorinated hydrocarbons we have again to turn to North American experience.

One classic case, which has been fully investigated, is that of Clear Lake in California. This has often been quoted, but it illustrates the case so well that I feel bound to deal with it again. This lake is used for recreation and fishing, but there have always been many serious complaints due to the clouds of a small gnat, a *Chaoborus*, whose larvae live in the water. This insect does not actually bite, but it occurs in such numbers that it is a serious pest. Holidaymakers also complain about the smaller numbers of biting mosquitoes and midges, which are a feature of many American inland holiday resorts. It is important to stress that there really was an insect problem; some accounts of Clear Lake give the impression that the authorities deliberately poisoned the area for no good reason. The next point to remember is that attempts were made to use insecticides which would have the least possible harmful side effects. After considering all the chemicals available, it was decided to use **DDD** (the same substance is also known as TDE), an insecticide related to DDT but shown under experimental conditions to be less lethal to fish. It was known, as mentioned above, that fish are very susceptible to DDT, and water containing 0·5 parts per million may be lethal. DDD concentrations of this level did not appear to harm most fish in preliminary tests. In the first instance, in 1949, DDD was applied at Clear Lake in an amount which, had the whole amount been dispersed through the total volume of the lake, concentrations in the region of 0·015 parts per million would have been obtained. As it was expected that most of the DDD would sink to the bottom, where the midge larvae were found, the exposure of the fish was assumed to be slight. In 1949 the operation was a spectacular success. The midges were almost completely eliminated, other invertebrates were certainly harmed, but not seriously and the populations seemed to build up rapidly. The fish, and other wild life, including fish-eating birds, seemed unharmed. In 1950, although no further control measures were introduced, the midges did not return in sufficient numbers to cause serious complaints. In the next two years the midges did increase in numbers up to approximately their original level, but control from one application of an insecticide had lasted for several years, and it was assumed that a safe and economical method of midge control had been found.

Further applications of DDD were made, and gave reasonable, though not always 100 per cent, midge control. However, by 1954 serious side effects were suspected. In fact, the first application of DDD appears to have seriously reduced the breeding of the Western Grebe in the lake area, but it was not until the winter of 1954 that large numbers of dead grebes were found and a public outcry occurred. The exact cause of these deaths took several years to find. It now seems that this was a striking case of the possible concentration of a poison in a food-chain, with death caused only to the animals at the end of the chain. As mentioned above, the amount of DDD originally added was only sufficient to give a very low concentration if equally dispersed on the water or the bottom mud. Even after several applications this amount is still low. However, all forms of life have concentrated the DDD. Plankton is found to contain about five parts per million. Small fish, which feed on plankton, contain about twice as much. The predatory fish which eat these small fish contain much higher amounts, but the greatest concentrations of DDD and other substances produced by its metabolism have been found in the grebes, which have as much as 1,600 parts per million in their visceral fat. We shall discuss the significance of small insecticide residues in birds and other animals later, but there is no doubt that large amounts like this are sufficient to cause fatal poisoning, even if lower residues may not always be harmful. The interesting fact to emerge at Clear Lake was that this concentration in food chains could occur, and could kill the fish-eating birds at the end of the chain. This has caused ecologists to worry in case similar effects should occur in many other ecosystems.

There have been many other reports from America of invertebrates, fish and fish-eating birds being killed when streams and lakes are treated, accidentally or on purpose, with insecticides. So far similar cases, with proved deaths of fish-eating birds, have not been found in Britain, even though, as will be seen later, our fish-eating birds do have higher levels of insecticide residues than other species with different feeding habits. We do have many proven cases of birds poisoned by insecticides in Britain, but so far aquatic food chains do not seem to have given levels *quite* high enough to do serious damage. Almost always streams in forests which have been sprayed from the air have been contaminated.

There is one well-documented British experiment which parallels the American experience at Clear Lake. This showed that DDD could control several species of midges with bottom-feeding larvae in similar amounts to those used in America. Carp in these ponds showed quite high insecticide residues, but survived and grew, though other experiments suggest that trout would have died. No data regarding birds are available. However, it seems clear that if we used insecticides in Britain as they were used at Clear Lake, equally serious results to wild life would occur. We are fortunate in not often having such insect problems in southern Britain; attempts to control midges in the Highlands of Scotland with persistent insecticides used over wide areas could be equally damaging. In America there have also been several instances where DDT and other insecticides, present in very low concentrations, have damaged fresh water molluscs. These animals not only obtain their oxygen from the water, and so need, like other aquatic animals, to "breathe" large volumes. They are also filter-feeders and obtain their food in minute suspended particles; this again brings them into contact with very large volumes. DDT at a concentration as low as 0·005 parts per million is taken up by some molluscs, which concentrate it in their bodies, producing toxic and even lethal concentrations. So far little damage to molluscs has been reported in Britain.

We have seen that in Britain, where DDT has not been used too frequently, it has seldom done much harm. However, there are some examples which show the danger of too frequent applications. Orchards in Britain tend to be more heavily sprayed with pesticides, both insecticides (against codling moth, winter moth, capsids, woolly aphids to name but a few pests) and fungicides (against scab and mildew), than any other areas in this country. Some fruit farmers are known to spray as many as twenty times in a season. We have already seen (p. 99) how copper fungicides can build up in the soil, and eliminate earthworms. DDT and other persistent insecticides have also been shown to accumulate, and must have a considerable effect on the fauna. However, the most striking effect of insecticides has been the appearance of a new pest. The tiny red spider mite was known on outdoor fruit trees, but was not considered an economic pest (spider mites have given trouble in glasshouses for a longer period, but this is a different story). Since the last war it has changed its status into that of a pest. The explanation seems to be that repeated spraying, particularly with DDT, has eliminated the slow-breeding predatory insects which controlled the mites. These creatures breed very rapidly, they are comparatively resistant to DDT, and there is even evidence that low concentrations of the insecticide actually stimulate hatching of mite eggs. I will deal later with biological control, including the use of predators against pests; the spider mite was, unknown to most fruit farmers, kept under by a process of biological control before DDT and other chemicals upset the balance.

Worms and insects which survive in the soil in British orchards are known to contain substantial amounts of DDT, and it is probable that birds are sometimes killed by eating such animals, as were the American Robins. However, there is little concrete evidence of widespread deaths, and it may be that here again Britain has been lucky. It seems probable that a comparatively small increase in pesticide use might have more serious results.

We do have a good deal of information about the effects on the insect fauna of the soil. Many workers have found that immediately after applying DDT to soil at a rate of one pound per acre, most insects in the superficial layers die. A month or two afterwards the minute Collembola, which feed on vegetable matter, increase in numbers, giving populations higher than before the insecticide was applied. This again is due to the absence of predators, in this case mostly soil mites. It may take years before the original equilibrium is restored. We do not know how important these changes in the balance of soil fauna are; it would be rash to accept them as entirely harmless.

DDT has been widely used in Britain and other European countries to control various domestic pests. House flies depend on suitable conditions for the larvae (maggots) which live on organic matter including that on refuse tips and around farm buildings. It is on farms that they have often been serious pests, and soon after the war farmers were delighted because DDT gave apparently complete control; a light deposit of the substance made treated walls lethal for months, and prevented reinfestation of byres and stables. DDT was also successfully used to control cockroaches, ants and bedbugs. It eliminated bedbugs from buildings which previously resisted treatment. The deposits of DDT in the crevices which these insects haunt meant that the minimum amounts of insecticide acted in just the right places. Unfortunately, in a few years, control, first of house flies and then of other species, was found to be much less effective. Tests showed that strains of insects had developed which were resistant to DDT; these sometimes took as much as a hundred times the normal amount of DDT before they were killed. At first many workers attributed this to an insecticide-resistant mutation which had taken place, but now we know that, though such mutations may have occurred, resistant strains of insects are usually produced by selection from the natural population. We always find a good deal of variation in the susceptibility of different members of a population of any insect to an insecticide. If the dose of insecticide is insufficient to kill the whole population, the most resistant are likely to survive, and if these breed among themselves, a new and more resistant population may arise. Sometimes we find a population with some degree of resistance, due to a particular gene. In heterozygous individuals the degree of resistance may be comparatively small, but after selection the chance of producing homozygous individuals, which may be much

more resistant, is increased.   Whatever the cause repeated exposure to an insecticide has frequently produced strains of insects
which are very difficult to control.

The first species of insects which developed resistance to DDT
were flies, mosquitoes and lice, all of medical importance. The
reason for this is that such species are more generally exposed to
hazards from insecticides. The medical entomologist is usually
set the task of eradicating an insect which is a pest or a vector of
disease. He believes that no human being should be in danger.
On the other hand, the agricultural entomologist seldom tries
to eradicate the insect; his job is to reduce infestation so that a
crop can be grown profitably. It often appears that it is quite
easy to kill the majority of insects, but very difficult to exterminate
them. Now where the policy is eradication, and a small population of resistant insects escapes, this may breed rapidly to fill the
space left by the unresistant part of the population which has
been wiped out. Where the agricultural entomologist only tries
to reduce the population below the level where severe economic
damage occurs, there is less risk of producing a highly-resistant
minority. Furthermore, most agricultural pests are widespread,
part of the population living on crops and part, and often the
greater part, living on other types of vegetation. This is illustrated
by such insects as the Frit Fly, which does substantial damage to
oat crops, but which also occurs in much greater numbers,
without doing much damage, on grass in pastures. Even if a few
resistant members of such a species survive on a crop treated
with insecticide, they are likely to be swamped by the large
population of untreated insects from outside the crop. I must
confess that when first we heard of strains of insects which were
resistant to insecticides, I expected this phenomenon to be
restricted to those of medical importance. I was wrong. The
most striking cases have certainly been among vectors of human
disease, but many agricultural pests have produced strains resistant
to a variety of insecticides.

The first reaction of those responsible for control when a
resistant strain appears is to step up the dose of the insecticide.
In fact, this often succeeds, but it may aggravate the situation,

encouraging the emergence of an even more resistant population. From the point of view of wild life, however, the situation is serious, for the higher dose increases the risk of environmental pollution. Up to now resistant strains of most pests have been controlled by using another insecticide. The genetics of resistance is a complicated subject. As a rule an insect resistant to one poison is almost as resistant to related chemicals (e.g. to two organo-phosphorus insecticides) but it remains susceptible to chemicals with fundamentally different chemical structures. A few insects have produced strains resistant to many different insecticides, and sometimes control has looked like a race between the chemist, producing a series of different insecticides, and the insect developing resistance to one after another! The situation is seldom quite so gloomy. With houseflies, for instance, if DDT is withdrawn for one or two years the level of resistance often falls, and the chemical can be used again with some success. This suggests that insecticide resistance may be linked, in some cases, with genes which do not always favour success, and the resistant individual may be less, and not more, vigorous than the general run of the population.

Pest insects have commonly developed resistance. Why then should not other forms of wild life, for instance, birds, such as the American Robin, which has suffered so much when elm trees have been treated with DDT? The answer is that resistance in birds is not impossible, and in fact instances have been reported, but the chance of selecting resistance comparable to that of insects is rather remote. Most pest insects lay thousands or at least hundreds of eggs, and have a number of generations a year. There is thus a huge number of individuals from which a small number of resistant insects may be selected. Birds lay few eggs, and breed slowly, and the chance of selecting resistance in any species is therefore much less. There is more chance of annihilation.

b: BHC

Benzene hexachloride – BHC – is another chlorinated hydro-carbon which is widely used as an insecticide and which came into use at the end of the 1939-45 war. Technical BHC is a mixture of several isomers, one of which, gamma BHC, is the most potent. This is commonly known as **Lindane**, particularly in the United States. BHC is a contact poison which penetrates insect cuticle. It withstands heat, and can be dispersed in smokes. Unlike DDT it is somewhat volatile, and in enclosed spaces this can en-sure the killing of insects which cannot easily be reached by sprays.

BHC has one disadvantage as an insecticide, in that it has a rather objectionable smell and taste. It easily taints certain foods, including root vegetables like potatoes and carrots, and fruit, particularly black-currants. The smell and the property of tainting food reduces its usefulness and also the risk to man and to wild life. BHC is a persistent insecticide, but it is less stable than DDT, and in general it seems less dangerous to wild life. It has been found particularly useful against wireworm, which is a common pest in ploughed-up grassland. BHC has been used as a seed-dressing, and has been used with cereal crops. Small amounts, which do little harm to the rest of the soil-fauna, are used in this way. Care must be taken to avoid using too great a dose of the insecticide, as it is somewhat phytotoxic, and too heavily dressed seed may not germinate properly.

Certain types of dressed seed, treated with insecticides, have caused the death of great numbers of birds (see p. 138). BHC has not been seriously incriminated. The Game Research Association has made extensive tests of the effects of grain dressed with BHC on pheasants. In some the birds seemed to find this distasteful, and even when starved they were reluctant to eat it. Pheasants could, however, be conditioned to eat the treated grain, by being given a diet in which the amount of pesticide was gradually increased. Even so the grain normally sown was relatively unpalatable, though larger amounts were dug up and eaten than were taken when offered in dishes; perhaps the soil masks the smell. The BHC had some effects on the pheasants, though no deaths were attributed to it. It delayed the date on which the birds came into lay, but it had little or no effects either on egg numbers or on hatching. Analyses showed that the birds and their eggs contained residues of BHC, but these disappeared quite quickly when no more was eaten, and the birds seemed quite normal when returned to an insecticide-free diet. The general conclusion was that, under field conditions, it was unlikely that game birds in Britain would often ingest a harmful dose of BHC.

Residues of BHC are widespread in birds in both Britain and North America, but only in amounts that are probably harmless (p. 148). Clearly if it is used on a large scale, for instance in mosquito control, as it is relatively non-selective, it will kill most insects in the treated area. It has proved useful against ticks and mites in tropical countries; these are sometimes difficult to control with DDT. In orchards, used against the red spider mite, it has also killed predators which otherwise help to control this pest, and results have therefore sometimes been disappointing. Caterpillars pick up small doses and pass these on to insectivorous birds, but this has not so far been shown to cause bird deaths. The general conclusion about BHC would seem to be that it has not proved particularly dangerous to wild life, perhaps because unpalatability acts as a warning and prevents dangerous amounts from being consumed. Were it to be sprayed in large quantities from aeroplanes it might well do much more harm.

c: DIELDRIN AND RELATED COMPOUNDS

Several insecticides belonging to the cyclodiene group of chlorinated hydrocarbons have been in use since about 1948, though they were not widely used in Britain until the middle nineteen fifties. Those which are most familiar are **aldrin, dieldrin,**

heptachlor, endosulfan and **endrin.** They are all poisonous to mammals, and to man, endrin being as toxic as the more poisonous organo-phosphorus insecticides, the others being more comparable with DDT. This group includes the insecticides which have done real harm to wild life; but for their use there would have been much less public outcry about toxic chemicals used in agriculture. Like other organo-chlorine insecticides, these substances are very soluble in fat, but practically insoluble in water. They can act as contact poisons, passing easily through the insect's cuticle, and they are also lethal if eaten.

These insecticides seemed at first to be remarkable substances, with almost all the qualities which scientists were seeking. Except for endrin, they were no more poisonous to mammals than DDT, and in 1948 few people thought this was dangerous to man, and practically no one had suggested that it was likely to harm wild life (except to kill insects within treated areas). They remained potent for long periods; in fact their persistence was greatly under-estimated, but even if it had been realised, this would still have been considered an advantage. A reason for the great persistence of some of these substances is that, instead of breaking down to give more or less inert substances, as do so many pesticides, these are transformed in the soil or in living tissues to equally or even more toxic chemicals. Thus heptachlor turns to heptachlor-epoxide, a stable substance considerably more poisonous than heptachlor itself. Aldrin disappears quite quickly, and early workers thought that this meant an end to its potency; however it was soon found that it was transformed into dieldrin, which could be traced unchanged in soil for ten years or more. It was only later found that dieldrin could be stored by living animals, mainly in the fat deposits, and that it remained fully toxic and able to cause poisoning when once more it was mobilised.

Aldrin, dieldrin, heptachlor, etc., were not, as has often been suggested, put on the market without extensive toxicity tests. They were tried out on mammals, rats in particular, and precautions to prevent damage, particularly to those applying the insecticides, to domestic stock, and to those eating treated crops, were devised. These precautions have in fact proved reasonably satisfactory, at any rate in Britain and other temperate countries. There have been very few proven cases of accidental poisoning of man. In tropical countries, operators applying dieldrin over wide areas in anti-malaria campaigns have occasionally been poisoned, though even here I doubt whether one death has occurred for every million people who have been protected from disease.

In Britain aldrin and dieldrin have been widely used in several ways. They effectively control wireworms and other soil pests, and to reduce costs of application they have been included with the fertilisers which are commonly applied even when no insecticide is needed. They proved particularly useful against the carrot fly; carrot growers still insist that no other substance is anything like as successful. Then they have been used as a seed-dressing on wheat; this gave excellent protection against the wheat bulb fly, whose larvae attack the young plants in early spring. A further most important use has been in sheep dips and sprays; one application of dieldrin in early summer has given satisfactory protection from fly strike (flies lay their eggs on soiled areas of fleece, the maggots emerge and feed on the living flesh of the sheep) for a whole season, substantially reducing costs of management to the farmer and removing a cause of great misery to the sheep. House fly populations which have become resistant to DDT were susceptible to dieldrin, at least for a period, for resistance to dieldrin also has developed in due course in some cases. However, the existence of new fly poisons enabled those engaged in control to "ring the changes" and at least reduce the nuisance.

As has already been said, toxicology is a difficult subject. Different species of animal differ in their susceptibility to different poisons, and it is at present impossible to foretell with certainty the effects of a substance on one species by trial using another. Furthermore, the same species may be more tolerant of a poison at one season of the year than at another, and the way in which the poison is administered affects the results. No insecticide can be tested against every species of bird and mammal which may be at risk. It is usual to-day to test the toxicity of new pesticides for acute toxicity (LD_{50}) against a mammal, a bird and a fish, and also to investigate long-term effects of chronic poisoning with small doses on several animals. Some experiments extend over several years, and generations of test animals, so fertility and breeding success is studied. The vast majority of substances which, in the laboratory, are found to be potential pesticides, are scrapped as a result of unsatisfactory tests. These screening procedures are being improved year by year, and it seems that since 1961 when ecologists were officially consulted in Britain there has been an improvement in the general situation, and no *new* insecticides which gravely endanger wild life have been put on the market (see p. 184).

When the toxicity of the cyclodiene group of insecticides was tested against mammals, with the exception of endrin they did not seem to be particularly dangerous. However, further tests, particularly with birds, have shown that we previously under-estimated the dangers. The table shows that dieldrin in particular is particularly dangerous to quails, and experiments with other species of birds confirms this result.

Toxicity of some chlorinated hydrocarbon insecticides to the rat and the quail

	LD_{50} as milligrams per kilo body weight	
	Rat	Quail
Endrin	10	under 10
Dieldrin	100	35
Heptachlor	130	125
Gamma BHC	125	200
DDT	113	500

The most spectacular incidents in which wild life was affected by pesticides occurred between 1956 and 1961. In the spring of each year large numbers of many species of birds were found dead, particularly in the eastern part of England. The total number of corpses was substantial. As many as 500 dead woodpigeons were picked up on one day under a roost. Dead pheasants and partridges were also found in large numbers. In one area of 1,480 acres of woodland, it was estimated that 5,668 woodpigeons, 118 stock doves, 59 rooks and 89 pheasants died. On the whole seed-eating birds suffered the most, but hawks and other predators were also found dead. The total death roll must have been even higher than the number of corpses suggested, because experiments have shown that even a skilled observer misses many dead birds when these are distributed over rough ground, and also many corpses disappear within a few days, having often been eaten by foxes or other animals.

It was not until 1961 that the mystery was cleared up. Many people had been investigating it. Naturalists throughout the country played their part, and among voluntary organisations the lead was taken by the British Trust for Ornithology, the Royal Society for the Protection of Birds and the Game Research Association. Scientists belonging to the Ministry of Agriculture, Fisheries and Food and to the Nature Conservancy, which was the first official body to set up a team of investigators, all co-operated. It became clear that the most serious incidents occurred in the eastern counties of England. These were the areas where cereal crops were mainly grown, and the use of dieldrin as a seed-dressing was clearly implicated. The situation was confused because almost all the seed had been treated with organo-mercury fungicides; though these can apparently be dangerous, all the evidence gained at this time suggested that though mercury residues were found in the birds, these were too low to be harmful (see p. 101). BHC used as seed-dressing was also exonerated in most instances (p. 133).

As an entomologist who realises that insect pests can do severe economic damage, I cannot help feeling that this is a very unfortunate story. Dieldrin was used as a seed-dressing, not with any sinister intent, as some writers have implied, but in the hope of getting the best result using the *smallest* amount possible of the insecticide, with minimal environmental contamination. As a rule only about three ounces of the dieldrin were stuck on to the wheat sown over a whole acre. The insecticide protected the young wheat plant, but had little effect on the soil fauna. When an insecticide is broadcast, more is used and many more soil animals are killed. It does not, however, have a direct effect on the birds. The trouble was that in the spring of 1961 (and no doubt in the earlier years also) the birds, and particularly the pigeons, deliberately dug up the treated grain. A pigeon can dig up as much as two ounces of seed corn in a day, and it will probably receive a lethal dose of dieldrin from this amount. Pigeons feeding mainly on buried grain will almost certainly eat enough in a few days to be fatally poisoned if it is dressed with dieldrin. Pheasants also can easily obtain a lethal dose.

The main cause of bird deaths was due to this actual digging up of poisoned grain. The weather in early 1961, at the time of sowing, was wet, and this meant that some of the grain was accidentally left on the surface. It is also alleged that some farmers, wishing to control woodpigeons which can be serious pests, deliberately left dressed grain in piles on the surface of the ground. This, incidentally, is not a new complaint. In the *Shooting Times* in 1884 the following passage appears:

The Danger of Dressed Wheat

"An unusual conviction was recorded at Peterborough Police Court on Thursday against a farmer named Robert Goodfellow. The prosecution was conducted by the police, through the defendant having scattered dressed wheat on an already drilled and harrowed field, thereby causing danger to life, as numerous dead birds – woodpigeons, crows and a magpie – had been discovered on the defendant's and adjoining land. The prosecution took the view that danger to human life might result from the continuance of such a practice."

I cannot be sure what dressing was used in 1884; it certainly was not dieldrin, but the principle was the same!

It is difficult to assess the long-term ecological importance of the kill of birds from dressed seed. Pigeons were the most noticeable victims, and as these are a pest this was welcomed by most farmers, not only the few who poisoned the birds by deliberately exposing grain on the surface. Unfortunately, although nearly ten per cent of the pigeons died, this had little permanent effect on the population. We know that a much larger number of pigeons dies from starvation every winter, but that this quick-breeding pest can breed up to a high level again in one season. Poisoning is not really an efficient means of control. Other birds, ones which are not economic pests, often breed much more slowly, and it is probable that had poisoning on the scale seen in 1961 gone on for several years, many species would have been wiped out, at least in the corn-growing areas. As will be shown later, long-term effects on predatory birds are still suspected. The dead birds, particularly the pigeons, were also a source of danger to other forms of wild life.

Animals killed by eating most poisons can themselves be eaten by carnivores without danger, but the birds poisoned by dieldrin remain toxic and were responsible for many deaths of other mammals and birds; this is no doubt because this insecticide is so very stable, and it retains its toxicity when passed from one animal to another. The victim which has been most carefully studied is the fox. During the winter of 1959-1960, and particularly in January and February, 1960, some 1,300 foxes were actually found dead. Sick animals were also observed. These lost their fear of man, and one actually wandered into the yard of the Master of a Hunt. The sick animals died in a few hours, suffering convulsions, coma and apparent blindness. At first many of those investigating the outbreak thought it was due to an epidemic of a virus disease, and some are not yet convinced that all deaths were due to dieldrin poisoning. I personally think that the case was proven. The area of high fox mortality was the same as that in which these seed-dressings were widely used. No infectious agent was found in the corpses; high residues of insecticides were found. In some trials pigeons which died after eating dressed grain were fed to foxes which in turn died. During the same period of 1959-60 farm dogs and cats, badgers and carnivorous birds were also found dead, and there is little doubt that they also had eaten birds killed by seed-dressings. These deaths occurred at the time or very soon after the catastrophic kill of seed-eating birds, when a fox, a cat or an owl consumed only one to two birds containing a high dose of poison. Sometimes a carnivore may have needed to eat several corpses to obtain a fatal amount, and to that extent we had some concentration of poisons in the food chain. The process, illustrated by the concentration of DDT in the Clear Lake affair, by which over a long period low concentrations of residues are gradually concentrated as they pass up a food chain, in Clear Lake from plankton to small fish to fish-eating birds, is rather different, and is dealt with in more detail on page 156.

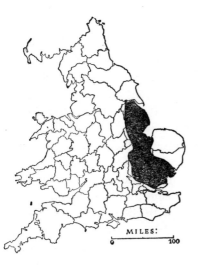

Fig. 6 Area of high fox mortality in 1959-1960.
(After D. K. Blackmore.)

MILES:
0   100

Dieldrin seed-dressing had been proved to be the most efficient protection against the wheat bulb fly, particularly in autumn sown wheat. This pest is commonest in eastern England. The female fly lays her eggs in late summer, but the larva, which damages young wheat plants, does not emerge and start feeding until early spring. If wheat is sown early in the autumn the plants may be big enough to survive by the time the wheat bulb fly larvae get to work. The worst damage is done to wheat sown late in the autumn. It is now possible to foretell whether an attack is likely. Samples of soil in wheatfields can be examined before sowing, and the wheat bulb fly eggs counted. If the estimated number per acre is under 1,000,000, no serious damage is likely. If more eggs are present, the seeds need protection.

Studies of woodpigeons and other birds have shown that they seldom dig up corn sown during the autumn. At that time there are other, and more easily reached, sources of food, including particularly grain accidentally dropped on the stubble during harvesting, acorns, beech mast and other seeds. Early in the new year, however, the food supply is much less, and more attacks are made on sown fields. When all the facts were understood, an attempt was made to remove the danger to wild life and yet give reasonable protection to the crops. Farmers in areas where wheat bulb fly was a serious pest had no doubt that seed-dressing with dieldrin was the best protection, and they wanted to continue its use, unless they were keen on shooting and game preservation when they wanted equally to save their pheasants. A compromise solution was found. It was agreed that dieldrin was the best protection for autumn sowings, and that these were the least danger to birds. A voluntary scheme was adopted by insecticide manufacturers, farmers and Government advisers, which allowed dieldrin seed-dressings in autumn in fields where high populations of wheat bulb fly eggs had been found. Spring sown wheat was not to be dressed with dieldrin, no matter how high the bulb fly egg count; control here can be got successfully with BHC. This voluntary ban has proved reasonably successful. No more mass slaughter of seed-eating birds has occurred in succeeding springs. Though this is a considerable advance, it is unfortunate that these substances appear to have other equally serious though less direct effects on many species of birds, so opposition to their use continues.

It has already been mentioned that dieldrin is a very efficient insecticide for controlling carrot fly, a pest which does substantial economic damage. Growers have found that one application of the insecticide to the soil gives a high degree of protection against the pest for as long as ten years. Twenty years ago most scientists and farmers would have accepted this as an unmixed blessing. We now see that such persistence can have disadvantages. The dieldrin kills not only the carrot fly, it kills many other soil insects. Beetles which are predators on other insects are among the victims. Under normal circumstances these predatory beetles consume many insects which are themselves pests; they are responsible for some degree of biological control (see p. 195). Thus when cabbages or cauliflowers are attacked by the grubs of the cabbage root fly, the attack may be most serious in soil previously used for carrots which have been protected by dieldrin. The predacious beetles are slow-breeding insects, and it takes them several years to recover their numbers. So far this process has not had any great economic importance to farmers, who can usually control cabbage root fly by other means, but as strains of this insect which are resistant to insecticides are emerging, the reduction of predators may in future be more serious. We are also concerned because the persistent organo-chlorines in soil may be responsible for further environmental contamination.

d. EFFECTS ON PREDATORY BIRDS

We have seen that organo-chlorine pesticides, and particularly dieldrin seed-dressings, have been responsible for mass killing of birds, and for the death of other birds and mammals feeding on their corpses. There is circumstantial evidence that these substances have also been responsible for serious damage to a number of other species of predatory birds, and that this damage is not restricted to corn-growing areas, though the situation there is most serious, as some evidence also comes from remote areas of the Scottish Highlands.

It is difficult to obtain accurate figures showing how the numbers of most wild birds have changed over the last fifty, or even twenty, years. It is even more difficult to obtain reliable data for insects. Nevertheless, unless we have such information we cannot assess the ecological effects of any environmental changes, whether these are due to agricultural practice, to industrial pollution or to the use of pesticides. Even when accurate statistics are available, it is sometimes difficult to be sure that we have always given the correct explanation of the cause of any change. Fortunately information has been collected for a few species of birds which makes it possible to advance a reasonable hypothesis.

There is no doubt whatever that there are fewer peregrine falcons and sparrow hawks in Britain in 1966 than there were in 1950, and, perhaps more serious, in some areas the few remaining birds are not breeding normally. The peregrine has been the most thoroughly studied. It is believed that its population was remarkably steady for several hundred years, up to 1939. The number of breeding pairs in the whole of Britain was about 650, and it seems seldom to have increased or decreased by more than about ten per cent. The birds chose nesting sites on cliffs, and it was the number of suitable sites, plus the number of surrounding territories, which controlled the population in most parts of Britain. During the war of 1939-45 rigorous control measures against the peregrine were taken in southern Britain, and numbers fell. When controls were stopped after the war, the peregrine made a remarkable recovery, and was approaching its previous level by the early 1950s. Since then the situation has altered, and the present position is critical. In 1962 in the remote areas of the Highlands of Scotland practically all territories were still occupied by breeding pairs, and more than forty per cent of these successfully reared young. At the other extreme in southern England of sixty-six territories occupied in 1939 no less than forty-six were untenanted. In the remaining twenty-two, no nesting was seen in the majority, and only three pairs reared young. There is a striking gradation from the north of the country to the south, both in numbers of birds and in breeding success.

The cause of damage to the peregrine could not be easily determined. Keepers shot a larger proportion before the war than after, yet pre-war shooting did not significantly affect the numbers nesting. Egg collectors were very active before 1939, when the majority of nests in the south were robbed; even this

Fig. 7 Peregrines in Britain in 1961. In each region the first figure shows
the percentage of territories occupied, the second (in box) the percentage in which young were reared. (After D. A. Ratcliffe.)

did not reduce numbers. Presumably any surplus stock migrated from other regions. No environmental changes could be correlated with the disappearance of these birds from so many territories.

When peregrine corpses were found, they were subjected to post-mortem examination; no signs of organic disease were seen. Analysis showed that most of them did contain substantial amounts of organo-chlorine residues, and in a number of cases the amounts were sufficient to have caused death. Other birds which were shot or killed by hitting moving vehicles contained rather lower insecticide residues, but still considerable amounts. Eggs which failed to hatch also contained pesticide residues. Now we know from experiments with other species of birds that breeding may be prevented by sub-lethal doses of organo-chlorine insecticides. These substances may cause sterility, so no eggs are laid. In other cases eggs have been laid, but do not develop properly. A third effect of pesticides can be to affect behaviour, so that eggs are not incubated and, in extreme cases, are broken by the parent birds. Peregrines feed on birds, particularly pigeons, which we know can pick up a substantial amount of pesticide. When we realise that the peregrine has been most severely affected in those parts of Britain where pesticides have been most used, the evidence that they are responsible for the decrease in numbers is almost overwhelming.

Similar studies are being made on other predatory birds. The sparrowhawk, which also feeds by killing living birds, was until recently a common breeding bird throughout Britain, including the eastern counties of England. It was so common that ornithologists did not bother to keep records, and a number was usually seen on every gamekeeper's gibbet. To-day it is a rare bird of passage in many areas. In 1965 only one breeding pair was reported from the whole of Norfolk; in 1949 there were several hundred pairs. Only in the more remote parts of Britain does the sparrowhawk still breed successfully.

Many of our other predators, including the kestrel and the tawny owl, have more catholic feeding habits, eating mammals and insects in considerable amounts. These foods are less likely than birds to contain large amounts of organo-chlorine residues. Although there has been some reduction in kestrel numbers, and this mainly in eastern England, the situation is nothing like as serious as with the sparrowhawk. Tawny owls appear to have been affected hardly at all. These results again support the hypothesis that bird-eating hawks are most likely to be damaged by pesticides.

A third species of predator which has probably been adversely affected is the golden eagle. Eagles may live for thirty years, and up to 1966 there has been no substantial decrease in their numbers in Scotland. However, while in central Scotland most pairs continued to rear young, breeding success in western Scotland is very poor. Thus of forty pairs studied from 1937-60, twenty-six (72 per cent) reared young successfully. As four nests were robbed, the success was even higher among the birds not so treated. In the period 1961-63, out of thirty-nine pairs only ten (29 per cent) reared young; only one nest was robbed. In the earlier period only one pair which was studied did not breed, and few nests contained broken eggs. In the period since 1961 sixteen of the thirty-nine pairs did not lay, and of those which did nearly half broke their eggs.

Unhatched eagle eggs have been analysed, and found to contain amounts of organo-chlorine insecticides which may be responsible for their death. We have no figures to show that the eagles which did not lay, or which did not look after their eggs properly, had higher amounts of pesticides in their tissues than the more successful birds from the central Highlands, but a study of the diets in the two regions makes this a distinct possibility. In the central Highlands eagles live mainly on grouse, which, feeding on heather, are unlikely to be seriously contaminated with pesticides. In the Western Highlands sheep carrion is an important part of the eagle's food. In recent years sheep have been dipped in preparations containing dieldrin, and it is known that this can seriously contaminate corpses. The circumstantial evidence that breeding success in eagles in the Western Highlands has been reduced by pesticides is therefore more convincing than that which has sometimes convinced a jury in a case of murder! In this case contamination up to now is not sufficient to kill the adult birds. It is hoped that the ban on dieldrin for sheep dips, which operates from 1966, will soon cause pesticide levels to fall and allow breeding to start again before it is too late.

e: THE SIGNIFICANCE OF RESIDUES IN BIRDS AND OTHER ANIMALS

As already mentioned, chemical analysis has shown that residues of chlorinated hydrocarbon insecticides can be found in a great many species of birds and other forms of wild life. It is not so simple to discover the biological significance of these residues. Many writers – and there is now an immense literature on the subject – assume that if a chemist can detect the presence of a trace of a poisonous substance, then that poison is damaging the animal in which it occurs. This is certainly by no means always the case. As chemical techniques improve, smaller and smaller amounts of some substances can be detected. The methods of estimating pesticides are now very accurate, and to-day we can record contamination which would previously have been undetected. Very low residues found in dead birds can be accurately determined; we can often be sure that they would not cause death. We can be practically certain that high residues are harmful, that they probably cause death if they reach certain levels, and at somewhat lower concentrations they would prevent normal reproduction. Intermediate levels of poisons are the most difficult to interpret.

Careful chemical analysis shows that not only do pesticides commonly occur in wild birds and mammals, but also that other substances usually thought of as poisons are normal constituents of living tissues. Thus arsenic is found in measurable quantities in all forms of life. Human blood contains about 0·64 parts per million (a concentration similar to the level of pesticides of equivalent toxicity often found in wild birds), fish muscle has as much as four parts per million of arsenic, and the human adrenal gland has been found to contain from 70 to 400 parts per million. Mercury is another poisonous substance universally found in living tissues; the human adrenal gland again contains from 2 to 160 parts per million. Both arsenic and mercury can be very poisonous when eaten, but the amounts of these substances found in normal living creatures are presumably not doing them any harm, and in fact they may be essential to life. It may be noted that the occurrence of mercury in normal animals has made post-mortem diagnosis of mercury poisoning very difficult, and arsenic poisoning may be equally difficult to establish.

Fortunately insecticides such as dieldrin or DDT are not normal constituents of living tissues, so if they are found this must be due to the animal somehow absorbing the pesticides. Until recently analytical methods were not able to recognise these chemicals in very low concentrations. The introduction of paper chromatography was a considerable advance, but limited identification to levels above 0·5 parts per million. More recently gasliquid chromatography has made it possible for the skilled operator to identify chlorinated hydrocarbons at much lower concentrations. The actual limits depend on the size and nature of the sample analysed. Levels in the region of 0·01 parts per million can be detected under favourable conditions when they occur in animal tissues, and even lower concentrations have been measured in water, where impurities do not interfere with the analysis. The biological significance of such results is what we have to discover.

In order to find how widespread insecticidal contamination is, birds and their eggs and also other animals obtained in different ways have been analysed. Various different things have been investigated. Wild birds and animals suspected of being poisoned have been examined to confirm or disprove the diagnosis. Various forms of wild life taken at random from the wild population have been examined to measure environmental contamination. Then birds and mammals have been fed diets containing known amounts of pesticides. This has enabled the LD_{50} to be determined. In some cases the effects of lower doses on breeding reproduction and behaviour have been studied. Finally analyses of the bodies of these experimental animals have been made, to relate the amounts of toxic substances in the various tissues to the doses administered and to the mortality and other effects.

Logically, the first step in elucidating this problem would have been to determine experimentally the toxicological properties of all the insecticides used. In fact, most of this work was only attempted when the problem had emerged, and much more remains to be done before we will fully understand the effects of many pesticides on birds and mammals. Early work was complicated by defects in experimental technique, and some results of analyses, particularly those where the amounts of toxicants were small, cannot be fully accepted to-day. Nevertheless we now have a reasonably clear picture of the effects of some pesticides, and I will deal with this side of the question first, and then, in the light of the findings described, I will try to explain the results obtained in the field from the study of wild birds and mammals.

One difficulty about the experimental approach is that many species of wild bird cannot easily be studied in captivity. Few of the hawks breed in confinement, and none of the British species can be obtained in sufficient numbers for experiments likely to give results which can stand up to statistical analysis. We know that different species differ greatly in their sensitivity to different poisons, so results obtained, for instance, using domestic fowls must not be applied to hawks without some caution.

DDT has been used in a great many experiments with domestic fowls, pigeons, quail, and other game birds, as well as with rats and other experimental animals. Lethal doses in various foods have been determined, and the bodies of birds and mammals dying of poisoning as well as controls and those receiving sub-lethal doses have been analysed. Early in this work it became apparent that the insecticide is not uniformly distributed through the tissues, and also that a considerable amount of DDT is apparently converted into other compounds. Some investigators have not always distinguished pp'-DDT, the chemical mainly used as an insecticide, from its metabolites, TDE and DDE, which make up substantial parts of the residues found post-mortem. It is not at all easy to determine the significance of these various compounds. It has been known for a long time that DDE is less toxic than DDT. Some workers have thought that DDE is almost non-poisonous while others have assumed it is equivalent to DDT. In fact it seems that DDE is probably between a half and a third times as toxic as DDT. DDE is almost entirely produced as a metabolite by animals which have ingested DDT. A store of DDE may mean that a dangerous dose of DDT has been ingested and then changed to DDE, or it may also mean that, in a predator, for instance, the animal analysed has itself ingested the less-harmful DDE which had already undergone this change within the prey. Analyses alone cannot show the origin of the DDE. TDE is used as an insecticide, and may thus have been ingested as such. It is, however, comparatively little used in Britain, and TDE found in analyses of corpses is usually due to the breakdown of DDT itself after death. It has been shown that this transformation may take place even at sub-zero temperatures when tissues are stored in deep freeze. In the soil, DDT remains unchanged as DDT, though it disappears over a period of years. DDE usually only appears in substantial amounts in tissues.

DDT (or DDE or TDE for that matter) is very unevenly distributed throughout an animal's body. All organo-chlorine insecticides are readily dissolved in fat, so it is not surprising that the highest concentrations are found in the adipose tissue. Under certain circumstances these considerable stores of poisonous substances appear to be comparatively harmless. Much lower concentrations in other tissues may have more damaging effects. Thus several species of birds and mammals, including man, have been found to have concentrations of DDT or its metabolites as high as 1,000 parts per million in the fat without showing any pathological symptoms, whereas residues of 100 parts per million in other tissues, including the brain, breast muscle, heart or liver only occur in cases where severe, and usually fatal, poisoning has taken place. It is generally agreed that concentrations of DDT in excess of thirty parts per million are only found in these tissues in animals suffering from some damage from the poison.

There is still some argument as to which organ should be analysed to give the most meaningful result. Some workers consider that pesticide levels in the brain are the most significant; unfortunately bird brains are small, and as they decompose soon after death they may be difficult to analyse. Breast muscles usually yield a larger sample, and the analyses seem to reflect reasonably the intake of insecticide and the exhibition of symptoms. On balance, however, analyses of liver, if only one tissue can be examined, are probably the most useful. If the purpose of the analysis is to find out whether there has been *any* exposure to pesticides, then this will probably be most easily detected by analysing the fat.

As has already been stated, DDT may be stored, apparently harmlessly, in the fat. There is some evidence that fat animals can consume more DDT than can thin members of the same species without showing symptoms of poisoning. The DDT remains in the fat for long periods; these may extend for years in the case of man. However, the stores are slowly depleted if no further DDT is consumed. DDT levels in muscle, liver and other tissues fall in a few days. Stores of DDT, and even of DDE, can be mobilised from the adipose tissue if an animal is starved and its fat deposits are used up. Thus sparrows and other birds have consumed considerable amounts of DDT without apparent harm, but when they have been given a much reduced amount of food (even if this is free from DDT) they have soon developed symptoms and have even died.

As is shown below, most wild birds that have been analysed in Britain have some DDE in their bodies, but in the majority of cases this reaches a level of less than one part per million (in, for instance, the liver). This is incontrovertible evidence that such birds have lived in an environment polluted with insecticide, but it also suggests that these birds have not in fact suffered any damage from the pollution. If amounts of this order of magnitude are found only in the fat, there is little doubt that it is harmless.

Feeding experiments with DDT have been done on breeding

birds. These have established that the insecticide can have harmful effects on fertility, on egg production, on the development of the chick in the egg, on its progress after hatching, and even on the maternal behaviour of the parent bird. Sterile eggs have been found with substantial amounts of DDT in the yolk. Other eggs which have not developed have had rather low amounts of DDT, but it seems that here the parent was affected rather than the egg. It is therefore unsafe to draw conclusions from low insecticide residues in eggs. It is difficult to generalise about the effects of insecticides on fertility, except that it seems that lower levels can upset breeding than give rise to tremors and other symptoms of poisoning.

Feeding experiments with dieldrin have shown that this substance is much more toxic, to birds, than is DDT (see table on p. 137). Tissue residues of dieldrin of ten parts per million or more are almost certainly harmful, and levels not much higher than this frequently seem to be lethal. Fertility has been reported to be seriously affected by as low concentrations as two parts per million.

It should be stressed that, though many of these experiments have been continued over periods of months, and occasionally of years, we have so far only investigated experimentally the acute toxic effects of pesticides. We have not yet got sufficient information to rule out the possibility that chronic effects of poisoning may eventually appear as a result of tissue levels lower than thirty parts per million for DDT and of ten parts per million for dieldrin. These are more likely to appear in long-lived animals like man, or perhaps in golden eagles which live for thirty years. There is less risk to small birds, which seldom live more than one or two years.

Care must be taken to distinguish between the results of analyses of random samples of birds, and of birds which are suspected of having been killed by being poisoned by insecticides. An attempt has been made to obtain a random sample by taking corpses killed by motor cars on our roads. It has been suggested that this may include some poisoned birds which were too sick to fly away in time. However, analyses of road casualties seem to agree fairly well with those of birds which have been shot, so they probably give a fairly accurate picture.

Fig. 8 gives the results of the examination of one such sample. It shows that the levels of insecticide residues differ considerably in various types of birds. Thus only low levels occur in vegetarian birds including those like moorhens which frequent fresh water. Omnivorous birds, and those feeding mainly on terrestrial invertebrates have slightly higher residues. Terrestrial predators have higher residues, those feeding mainly on mammals, like tawny owls, containing less insecticide than species like sparrowhawks which feed almost entirely on birds. Fresh-water birds, such as herons, which feed on fish, have very high residues.

Analyses have also been performed on sea-birds. Most naturalists expected these to be relatively free from contamination with insecticides, but in fact substantial residues were found. Sea-birds which feed on invertebrates are on the whole less heavily contaminated than those which feed on large fish.

It should be stressed that few if any of the birds whose analyses contributed to Fig. 8 contained sufficient insecticide for us to be able to say with any certainty that they had been seriously harmed. Even the herons, with some thirteen parts per million of insecticide residues, were mainly contaminated with DDE, which is probably harmless at that level. The importance of this type of investigation is that it shows how widespread insecticide contamination is, and indicates that a small increase in this general level could have a disastrous effect.

Analyses of birds known to have been poisoned with insecticides show far higher amounts of insecticides than those indicated in Fig 8. Birds which have clearly been poisoned fall into two distinct groups, as indicated above (p. 138 and p. 143). Pigeons and other seed-eaters, feeding on dressed seed corn, have contained substantial amounts of dieldrin, and the corpses of raptorial birds have been found with fifty parts per million or even more; this has been derived, at second hand, from their prey. An important point emerges from these British results. Only very few birds seem to have high enough residues of DDT, DDE or BHC for them to have been seriously damaged. On the

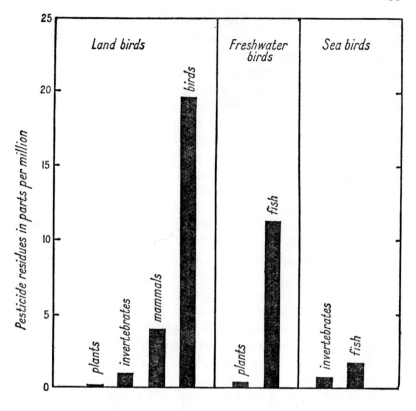

Fig. 8 Pesticide residues in birds from land, freshwater and the sea, related to the main types of food which they eat. The food is indicated by vertical printing—"plants", "birds", etc. (After N. W. Moore and C. H. Walker.)

other hand, dieldrin (and aldrin and heptachlor), which is so much more poisonous to birds, has quite often been found at levels which are certainly harmful and are probably lethal. I am of the opinion that very little damage has been done to terrestrial wild life in Britain by DDT or BHC, and provided their use is restricted, these compounds are nothing like as dangerous as dieldrin.

The origin of these insecticide residues in bird tissues provides an interesting problem. The chemicals must have been eaten, but we cannot always see how it got into the diet. Dressed corn is an obvious source, but we do not yet know how an insecticide sprayed on a tree or incorporated into the soil finds its way into a bird's body. Mention has already been made of food-chains, and of the process by which a chemical can be concentrated as it passes along such a chain. The classic example, which has been most carefully investigated, is that of Clear Lake in the United States, where the grebes, the final link in the chain, suffered the greatest damage (p. 128). A similar process is usually assumed to be operating generally. Although this is clearly happening, I am not sure that the food-chain hypothesis as usually understood explains the way in which insecticide damage to birds operates in Britain.

This is not to imply that concentration in food chains does not occur; it obviously does, and this explains why predators generally contain higher residues than their prey. Nevertheless, even with species like the sparrowhawk, which have almost certainly been severely damaged by pesticides, two distinct processes seem to be operating. DDT (or more commonly its metabolite DDE) seems to be concentrated in the commonly understood way. The result of this, as indicated above, is seldom if ever to yield a level which is actually harmful. On the other hand, we find a number of sparrowhawks, and birds of other species, which exhibit the symptoms of acute poisoning, and which contain very high residues, particularly of dieldrin. There is evidence to suggest that these hawks have been so unfortunate as to catch one, or possibly more than one, individual bird which has itself taken a very high dose of dieldrin. A pigeon, for instance, suffering from acute dieldrin poisoning becomes uncoordinated and flaps about violently for some time. Such damaged birds would be easily caught, and experiments have shown that tethered pigeons behaving somewhat similarly are particularly attractive to many hawks. If this is true it means that general environmental pollution at a low level may be less dangerous than the occasional opportunity (given, for instance, by the existence of even small amounts of dieldrin-dressed seed) for an individual bird to suffer acute poisoning. Of course, general pollution at *higher* levels, as has occurred in America when large areas have been sprayed from the air, even with relatively non-toxic insecticides, is equally dangerous.

There are special problems in aquatic environments. Fish have been thought to be particularly susceptible to poisoning, and have been killed when placed in water containing as little as 0·005 parts per million of BHC and dieldrin. Similar levels of insecticides have killed aquatic molluscs and crustaceans. Man would certainly not be harmed by drinking such water, and in fact food or drink with a thousand times as much pesticide would be harmless. The reason is not that the tissues of these aquatic animals are harmed at very low insecticide levels, but that the animals absorb the insecticides so efficiently and the tissue levels rise to ten parts per million or higher. There is a point where dilution is sufficient to prevent absorption; a fish may gain pesticide at 0·005 parts per million, and lose it at 0·001 parts per million. As was pointed out in the chapter on water pollution (p. 49) aquatic animals have to "breathe" very large amounts of water to get enough oxygen, and in this process they take up the pesticide when the level is high enough. Much of the concentration by the fish in Clear Lake was probably by this process, rather than by eating contaminated plankton. Thus very small amounts of pesticides in water will tend to be taken up by fish, and handed on to fish-eating birds. It is fortunate that the heron is apparently rather tolerant of insecticides in its tissues. So far in Britain, although a number of poisoned herons has been found, the population of the species does not seem to have been adversely affected. It was greatly reduced by the abnormally cold winter of 1962/63, but, notwithstanding the level of insecticides in most individuals, recovery since 1963 has been very satisfactory. Here again the heron has been fortunate; it is almost certain that had the insecticide level been only a little higher, the effect on them would have been catastrophic.

Man also contains pesticide residues. Analyses in the United States have shown that, in human fat, about twelve parts per million of DDE is the average burden. The level rose from about two parts per million in 1950 to twelve in 1962; since then there has been no further rise, and there is some evidence of a slight drop. In Britain the average level is considerably lower, about two parts per million. This difference gave rise to the suggestion that in cannibal circles Americans would be no longer on the menu; less dangerous Britons would be preferred. In both Britain and America the insecticide levels in other human tissues are usually below one part per million. Dieldrin residues in Britain are apparently higher than in America, but they again are still below the levels which we know to be harmful to adults. Nevertheless the amounts of pesticides in man are not negligible. A well-covered American business man is likely to be carrying at least 100 mgs. of DDT and its metabolites. Although experimental subjects have, without apparent harm, tolerated at least ten times as much, more work on the chronic effects of such residues is still needed. The levels in human milk sometimes approach those likely to be harmful to babies exclusively breast fed, so the need for control is already apparent.

f: IS THE WHOLE WORLD POLLUTED WITH CHLORINATED HYDROCARBON INSECTICIDES?

There have been reports of scientists finding DDT and other insecticides in all sorts of situations where pollution seemed unlikely to have occurred. Small amounts of pesticides have been found in rain-water falling out of a clear sky, and in fish from remote parts of Antarctica. It is probable that sufficiently accurate methods of analysis could detect these substances in all parts of the globe. The question we must ask is whether this matters, and whether this general pollution has any biological effect.

Rain-water carefully collected with rigorous precautions to avoid further pollution in a rural area of the English midlands has been shown to contain appreciable amounts of BHC, the maximum figure being 0·0001 parts per million. Very much smaller amounts of dieldrin were recovered, and traces of DDT were possibly present, though so small as to be at the limit of the range of the analytical technique. The higher concentration of BHC is explained by the volatility of the compound. Another investigation, in Central London, gave similar results for BHC, but also showed dieldrin and DDT at higher levels. One sample had 0·0004 parts per million DDT. Detectable amounts of pesticides were also found, apparently associated with particulate matter, in samples of air. It has disturbed many people to find these substances in both air and rain-water, but it should be remembered that the levels are so low that only a few milligrams of pesticide would be added by rain to every acre of farmland, less than a twentieth of a milligram could be detected in the air breathed during twenty-four hours, and it is most unlikely that there would be any effect on the fauna. It would be wrong to assume that any of this insecticide in the air is necessarily absorbed by the lungs; it is possible that in some cases insecticide is lost to the body rather than gained in this way, for the level in the tissues may be higher than that in the air. We know that local effects are caused by volatile insecticides being carried, like herbicide spray, to adjacent fields, but in general this form of pollution of rainfall and air in Britain is probably at present biologically negligible.

Insecticide pollution has been reported from both the Arctic and the Antarctic. In the arctic tundra of Alaska, insecticide fall-out has been compared to atomic fall-out; small amounts of both radio-active isotopes and insecticides have been detected in lichens and other plants. The quantities of insecticides, measured in thousandths of a part per million, are most unlikely to be biologically important.

There have been several reports from the Antarctic. In the first, fish, seals, birds and snow from remote areas were most carefully analysed. The lowest limit of the analytical technique used was 0·0005 parts per million, so levels below this would not be detected. All the samples of snow were negative. On the other hand, fish contained detectable residues, the highest (in fish fat) being 0·44 p.p.m. The fat of an Adelie penguin and of a Weddell Seal had insecticide levels about half this value, and a skua yielded 2·8 p.p.m. of DDE. These levels are of the same order of magnitude as those reported from Britain. A further report concerned Adelie penguins and seals. All contained traces of DDT (and its metabolites). The amounts, however, are minute. Calculation shows that if all the ten-million odd Adelie penguins in the Antarctic contain similar quantities, there will be about half a pound of DDT distributed through the whole population. If penguins and seals are typical, the whole amount of DDT and metabolites in all the biomass of birds and mammals in the Antarctic may be less than the total applied to a single elm grove in a North American suburb.

It is difficult to account satisfactorily for this Antarctic contamination. Some people have suggested that these animals have concentrated the universally present insecticides in their environment, and that therefore the whole globe must be as highly contaminated as Britain and the agricultural areas of North America. This is impossible. Man has probably not yet produced more than a million tons of DDT. A great deal of this has been broken down, much remains in the soil, and even in the fat deposits of both Americans and Europeans. If the whole of the million tons had been mixed into the sea, the level of contamination would be in the region of 0·000001 parts per million; no living organism has been able to extract DDT from anything like as weak a preparation. As only a small fraction, perhaps less than a tenth, of the million tons of DDT which has been synthesised could possibly have been dispersed in the sea, we must seek some other explanation than the universal dispersion of DDT for residues found in the Antarctic animals. One suggestion is that U.S. bases or ships within a few hundred miles of the site may have discharged quite large amounts of insecticide. Another is that those of the birds and animals which are mobile picked up their residues in other parts of the world; the skua for one had probably visited many more populous places. I personally think that these Antarctic findings are very interesting, but that some explanation other than really serious universal pollution will eventually be found. Such pollution, at very low levels, does occur, but so far it has probably not done much damage. We should be warned by it, and by the way pesticides can apparently

11 Treatment of roadside verges with herbicides. *Above*, three months after spraying. The treated area, near the road, is lawn-like and does not obstruct the motorist's vision. Untreated areas near the hedge are unaffected and wild flowers flourish. *Below*, another small area experimentally treated with herbicide, showing how effects are restricted to the parts which are sprayed

12 Effects of excessive use of copper fungicide on orchard soil. *Above left*, soil profile in unsprayed orchard showing a good soil structure; *right*, soil profile in sprayed orchard. The soil structure is poor, and a mat of undigested vegetable matter remains on the surface. *Below*, earthworms from an area of 4 sq. ft. of the unsprayed orchard. None was found in the sprayed orchard

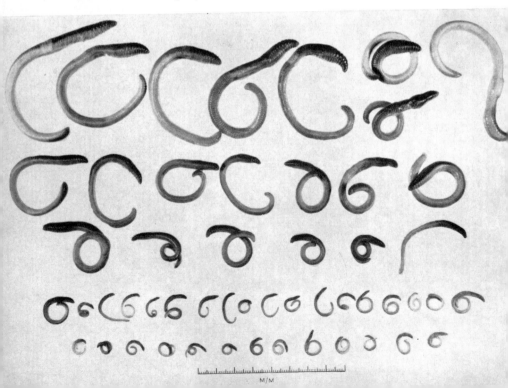

M/M

be transported throughout the world. Pockets of high concentrations seem to develop in unexpected places, depending on meteorological factors, ocean currents and the activities of man. The need for caution in the use of persistent chemicals is once again the main lesson that should be learned. The fact that more damage has not yet occurred in many places is no reason for complacency.

CHAPTER NINE

SUBSTANCES USED TO CONTROL
OTHER INVERTEBRATE PESTS

a: NEMATOCIDES

Minute nematode worms, on an average about one millimetre (about a twenty-fifth of an inch) long, form an important element in the soil fauna. Most of these nematodes feed on decaying vegetable matter, and play their part in nutrient cycles in the soil. They are thus beneficial to agriculture. There are also many species which, for at least part of their lives, are parasitic on plants, and these include a number of major pests attacking potatoes, cereals and most other crops. There are also nematodes which are vectors of virus disease in raspberries and other plants. The economic damage caused by eelworms, directly by destroying crops or reducing yields, and indirectly by making it impossible to grow what would otherwise be the most lucrative crop, is probably greater than that caused by all insect pests combined.

It is our agricultural practices which have allowed nematodes to become serious pests. Large areas of the same crop, grown year after year in the same ground, allow a low rate of nematode infestation to increase into a heavy one. This has been particularly noticeable with potatoes, where the potato root eelworm is distributed throughout the potato-growing areas of Britain, and where yields decrease seriously if the crop is repeated in several successive years. Parallel problems occur with sugar beet, attacked by its own eelworm, in cereals, and in a great many other crops.

A number of effective nematocides has been discovered. Dazomet and "D-D Mixture" kill eelworms and their usually resistant cysts, and soil so treated will grow susceptible crops successfully. These substances are poisonous to other forms of

162

life, including vertebrates, and they are also phytotoxic, so if widely used they might have serious harmful side effects. Fortunately for wild life, if unfortunately for the farmer, they are too expensive to use on field crops, though they are useful in glasshouses where high yields of expensive species are grown. I know of no reports of any serious pollution from such treatment.

So far the only way to reduce eelworm attack on the farm is by spacing crops of the same sort several seasons apart; this prevents a high eelworm population from building up. This crop rotation is probably beneficial to wild life generally, but as care must be taken to eliminate weeds which can maintain species of eelworms when susceptible crops are not grown there is always the risk of over-application of herbicides. Considerable progress has been made in developing races of crop plants which are resistant to eelworms; that is they become infected but show few symptoms of damage. Seed, which can transmit infestation, can be freed by careful heat treatment, as can bulbs and glasshouse plants like chrysanthemums. So far it would seem, then, that eelworm control does little if any harm to wild life. Were nematocides cheaper and more effective they might do much more harm! The eelworm problem is so serious that research to find such substances is being actively pursued, so the possible harmful effects of nematocides must be kept under careful scrutiny.

b: MOLLUSCICIDES

Slugs are serious pests in gardens and in agriculture. They do considerable damage to potatoes, where though they do not reduce the weight of the crop greatly their damage may make many tubers unsaleable. In the autumn they attack newly-sown wheat, and their consumption of grain may be so great that re-seeding is necessary. In a wet autumn slug damage may cost British farmers up to a million pounds, so there is need for control.

Metaldehyde ("Meta" fuel) is probably the best-known and most effective slug poison. Mixed with bran or some other suitable bait, or incorporated in pellets of similar substances, it is readily eaten. It appears to immobilise the slugs, which remain on the surface of the soil and die of desiccation. In gardens and fields, when such baits are spread, huge numbers of slugs are killed. However, this has seldom if ever been shown to have a permanent effect in reducing the slug population of a whole field or garden; as a rule the majority of the animals survive and quickly replace the casualties. Control of limited areas of a garden or glasshouse is possible, particularly if all decaying vegetable matter, which harbours slugs, is removed, and if the bait is spread in great quantities, but metaldehyde is not cheap, and slug poisoning in the field would seldom be economic even if it were effective. Metaldehyde is poisonous to vertebrates as well as to molluscs, and so is potentially a risk to wild life, but it is not sufficiently widely used to have done any appreciable damage. More efficient and cheaper molluscicides might be more dangerous.

Copper sulphate has also been recommended as a means of killing slugs. A solution sprayed on to the soil surface on warm damp summer nights, when slugs are most active, kills numbers of the animals, as well as damaging plants and other members of the soil fauna. The side effects are similar to those obtained when copper fungicides are used (p. 98). I have never known a slug population being controlled satisfactorily with copper sulphate.

Copper sulphate is also used against the liver fluke. This is a worm parasitic in sheep; it only exists where one stage in its life can find a water snail (*Limnea truncatulata*) in which to develop. The sheep pick up the worms from vegetation adjacent to water containing the snails. Infection can be avoided if the sheep are kept away from marshy ground with snail-infested pools. The snails may also be eliminated if twenty or thirty pounds of copper sulphate is applied per acre of marsh. This has serious effects on invertebrates, fish, amphibians and the vegetation. Streams are likely to be contaminated with quite high amounts of copper, again with serious effects on animal and plant life. Fortunately this treatment is comparatively seldom used, as it could have serious side effects.

CHAPTER TEN

THE CONTROL OF VERTEBRATE PESTS

Until 1939 or even later many wild vertebrates, both birds and mammals, were commonly spoken of as "vermin," and no methods of destruction were considered unsuitable. Gin traps, which caused terrible suffering to prey caught by a leg and not killed, were commonly used. All birds of prey were thought of as a menace to game, and were shot, poisoned or trapped unmercifully – in many cases without any harmful effect on the population of the species so persecuted. Control was most inefficient, and often wrongly directed. The real pests of agriculture, like rabbits, were actually encouraged in some parts of the country. Rats and mice, which do an immense amount of damage, were often ignored. We now tend to be somewhat more discriminating. Hawks and other raptorial birds, which the old-fashioned gamekeeper considered the worst type of vermin, are protected by law, and although some old-fashioned keepers still shoot hawks on sight, most of those interested in game preservation realise their virtues. We have learned that though hawks and owls may take a certain number of game birds, they live largely on real pests like rats which compete with pheasants and partridges for food. We have banned the gin trap, partly because it was inhuman, partly because it did little to control most pests. We are at least beginning to discover what are really pests, and to devise scientific methods to control them.

The control of bird and mammal pests illustrates the difficulties that may arise, particularly in densely populated countries, if this is to be efficiently done. Substances which kill one mammal are usually poisonous to another, so domestic pets and agricultural stock may easily be endangered. The same difficulty to distinguish between harmful and beneficial insects also exists, but does not cause so much concern. The death of a few ladybird beetles does not have the same impact as the poisoning of a pet dog or a herd of cattle, though the effect on the pest situation may be greater. Some progress in finding really selective poisons, killing only one species of pest, is being made, but we have a long way to go. Then it must be recognised that most vertebrate pests, except perhaps rats, are considered to be part of our wild life, and therefore worthy of preservation, by many people. Some animals, for instance deer, are in fact only pests when they get too common. Here pest control becomes population control; perhaps man himself can learn from this!

As was mentioned in the introduction, the **rabbit** is considered by many naturalists not to be a truly native British mammal. It was probably introduced by the Normans in the twelfth century, and though it was looked upon as a useful game animal, it did not become really common until towards the end of the nineteenth century. Rabbits then became very numerous in most parts of England and Wales. The extent of the damage they did to agriculture was not fully realised, or more effective control measures would have been instigated. Farmers in Wales in particular sold rabbits in great numbers, and some thought that they were more important than their ordinary crops.

Until 1953 those who tried to control rabbits used a number of methods. Snares and gin traps caught many of the animals. Rabbits were shot, often when driven from their burrows by ferrets. They were also netted with the aid of ferrets. They were harried by terriers, and attempts made to dig out their warrens. They were poisoned by gas pumped into the tunnels. If these methods were very thoroughly applied on one farm, they were sometimes temporarily successful, but within a short time, often within a year, re-invasion had taken place and the situation was as bad as ever. Farmers accepted rabbit damage as a fact of life, some tried to keep it to a minimum, many gave up the struggle. And, as stated above, some farmers even encouraged rabbits.

In 1953 myxomatosis reached England, and transformed the situation. Myxomatosis is a virus disease, endemic in South America. In its native area it is not particularly virulent among the indigenous rabbits. In South America the virus is carried from an animal suffering from the disease to a new victim by a mosquito. The Australians tried to use myxomatosis to control their rabbit pests. Rabbits had been deliberately introduced into Australia from Britain in the nineteenth century. Several of the first attempts were unsuccessful, but eventually the animals established themselves. They then bred very rapidly, spread all over the country, and clearly showed themselves to be very important pests. They ate down the grass needed for sheep, and as each adult rabbit eats about half as much as a sheep, and as sheep are so valuable, no Australian doubted that they were a menace. Several attempts to introduce myxomatosis failed, but eventually in 1950 this succeeded, and the results were dramatic. The disease had a mortality rate of very nearly 100 per cent. No other animals seemed susceptible. The rabbits were almost completely wiped out, and the stock-carrying capacity of the country was greatly increased. Farmers found that they had under- rather than over-estimated the amount of damage done by rabbits. In later years the disease, though still killing most rabbits, has proved less effective. The strain of the virus is less lethal, and some immune rabbits survive and breed. However, the demonstration of the damage the rabbits do, and the improvement achieved by control, ensure that the Australian farmers will apply other measures rigorously and supplement the continuing effects of myxomatosis.

Attempts were made before 1939 to introduce myxomatosis into Britain, but they failed. The disease appeared, apparently without the help of man, in the autumn of 1953. It had been deliberately introduced into France, and it is probable that a mosquito infected with the virus was blown across the Channel to start the outbreak in Kent. Myxomatosis is a disgusting disease, the appearance of the rabbits so affected shocked many Britons, and attempts were made to stamp out the first outbreak. These failed, the disease spread, sometimes encouraged by unsentimental farmers who deliberately transported an infected rabbit to another part of the country. In England it was found that the vector of the virus was the rabbit flea, so the disease might be expected to spread more slowly than in countries like France and Australia where it was transmitted by flying mosquitoes. Man may have helped its spread, though not its introduction, in Britain; by 1955 the disease had covered almost all the country, and had reduced the rabbit population by at least ninety per cent. In many of the areas where rabbits were formerly most common not a single specimen could be found. Many farmers were astounded by the result, as they realised, for the first time, just how much damage the rabbits had done. It is estimated that before 1953 they cost the country at least £50,000,000 a year. At last a really successful method of control had been discovered. However, we did our best to avoid using it. We could not stop myxomatosis or cure the rabbits, but we made it illegal to spread it deliberately, though a "black market" in infected rabbits still exists, and farmers on whose land rabbits reappear often risk prosecution by collecting an infected rabbit from a distance; up to 1966 there had in fact been no prosecution. The disease in Britain is now, as in Australia, somewhat less efficient, but the rabbit population is still very low as compared with 1953. Moreover, farmers now realise that control is important and use permitted methods with enthusiasm when they are required. Rabbit clearance societies flourish in most areas to co-ordinate their efforts. Myxomatosis was a "pesticide" which was almost completely specific. A few hares have been infected but otherwise all species of mammal save the rabbits have been completely immune. Except that it appeared to cause suffering to a cuddly mammal, it was the perfect control mechanism. If only rabbits had been invertebrates, about which we are not so sentimental, it would have been welcomed as the perfect pesticide.

Though myxomatosis did not kill other animals, and though infected corpses could be safely eaten by predators, it was responsible for important ecological effects. I have said that rabbits were not indigenous to Britain, but they have been with us for nearly 900 years, and they played an important part in many ecological processes. They were an important food source to carnivorous birds and mammals. When they disappeared, many naturalists feared that our predators would suffer. The effects were less than expected. Buzzards in Wales had lived very largely on rabbits. For the first two years after the rabbits disappeared, the buzzards seemed hard-pressed, but the numbers of adults remained more or less constant, and none seemed to die of starvation. However, hardly any of them bred during this period, and many people feared they would not survive. It seems that voles increased in numbers after that, occupying at least part of the niche vacated by the rabbits, and these provided sufficient food to allow the buzzards to rear young. No permanent fall in buzzard numbers has occurred; in fact these birds have increased slightly in some areas. Kites also failed to breed in the early years of myxomatosis, but resumed breeding at the same time as did the buzzards. There was also a fear that foxes would starve, or perhaps turn to an increased extent on to domestic poultry. There have been reports in some areas of more frequent raids on hen-houses, but not to any very serious extent. Foxes have not become less common except temporarily in areas where they were poisoned with dieldrin (p. 140). Research has shown that the importance of rabbit in their diet was over-estimated, and that small mammals, insects, carrion and even plant material may be their main foods. Other predatory birds have indeed become scarcer since the advent of myxomatosis, but, as we have seen (p. 143), this is due to other causes.

The most striking ecological effect of the disappearance of rabbits has been on the vegetation. Many areas of grassland, particularly on the chalk downs, were grazed hard by rabbits, particularly in areas where sheep farming had gone out of vogue. For the first two years without rabbits, the results were looked upon by naturalists as beneficial. Many of the interesting and beautiful chalk plants flowered profusely, as they had not done at least for many years. After this the result was less pleasing. It appeared that the rabbits had, by their grazing, prevented the hawthorn and other scrub plants from developing as well as keeping down the grass. To-day many former areas of chalk grassland have reverted to scrub, and the flowering plants previously considered typical of the habitat have been crowded out. For this reason some naturalists, though few farmers, regret the disappearance of the rabbit. They would like to see it return, not as a pest, but in sufficient numbers to reduce scrub growth.

When rabbits reappear, even in small numbers, they are, as I have said, strenuously attacked by man. The gin trap cannot now be used. Snares are set; these, if properly used, generally strangle the animals quickly and reduce suffering to a minimum. Snaring and shooting kills rabbits, it does not often harm other species, but it seldom controls the pest. The recommended technique to-day is to gas the warrens with cyanide. With care this is effective, and has no serious side effects unless badgers or foxes occupy adjacent tunnels. Some farmers, deliberately but illegally, kill these other mammals, but as a rule gassing has little harmful effect on wild life other than rabbits. No persistent residues are left after the treatment is concluded.

There is one point about rabbits which should be mentioned. It has been said that since myxomatosis the surviving rabbits have changed their habits, and live on the surface instead of in subterranean burrows. This does not seem to be the case, and there is no evidence of selection of a race with different habits. When young rabbits leave their mother they disperse, and it is these young animals which are found on the surface in newly invaded regions. After a time these animals make normal burrows just in the same way as did earlier generations. Young migrating rabbits have always lived on the surface, some have then moved into existing burrows, but when these are not available they have always dug in at the appropriate time.

Deer of various species, some indigenous (red deer, roe deer), others introduced (fallow, muntjac, sika and others), are much commoner even in agricultural regions than most people realise. They are seldom seen, and may do little damage, so they are usually encouraged or at least tolerated. In certain regions deer shooting is economically important. However, when deer populations increase, and this happens locally, they do much harm to crops and to young trees. Damage may be reduced by the expensive method of erecting deer-proof fences, but the only satisfactory course is to control the population below the level where damage is serious. This means skilful culling, and this can be made part of a sporting programme and so to pay for itself. Control by paid deer stalkers of the necessary skill is in fact more efficient, but is generally prohibitively expensive.

We also have ambivalent attitudes to **seals.** The grey seal is a pest to the salmon fishers of the Tweed, but a desirable member of our fauna to the rest of the population. The common seal is suspected of damage to fishing by some, is a source of profitable week-end shooting to others, and is thought of as a rare (because it is seldom seen – it is common enough in many places) and interesting animal by town dwellers. There is good evidence that grey seals do little economic damage unless their numbers increase too much, and attempts (which have aroused much public controversy) are being made to safeguard the species and the fisherman by a scientific policy of culling so that the population is controlled at the proper level. Eventually we will have to accept a policy like this for all our larger forms of wild life, not only in Britain but throughout the world.

Moles are looked upon as pests when they reduce the value of grassland by covering a substantial part of the surface area with their soil heaps, which destroy much of the turf and encourage weeds; the value of an area of pasture as grazing can be reduced by as much as fifty per cent, and the heaps can damage hay-making machinery. Moles also invade arable land, and sometimes their tunnels make it necessary to resow cereals. Some people object to moles because they eat so many worms, which are considered to be useful animals; there is no evidence that moles ever substantially reduce the worm population. On the other hand, moles are sometimes considered to be beneficial, as their tunnels may help to drain the soil. They are in fact an interesting form of wild life, perhaps sometimes a minor nuisance, but not a serious economic pest, and most people would be sorry if they were **exterminated.** Like so many of our indigenous animals, they are primarily woodland dwellers, which have found it possible to invade man-made open farmland and flourish there. In forests they do no harm whatever, and are usually left alone.

Moles were previously trapped more for the economic value of their skins than to eliminate them from farmland. There is now little market for the skins, and skilled trappers are becoming rare. It is not difficult to trap all the moles in a restricted area; even when hills are numerous there are seldom more than two or so moles to an acre, and new invasion of a garden, with no existing tunnels, by a single animal can wreak havoc until it is captured. The old-fashioned pincer type of trap is effective. It kills moles instantly when properly set, but seldom catches other animals. Occasionally a weasel is trapped, but may be small enough to squeeze out and escape. Otherwise, I have never seen anything but moles in mole traps. Trapping or other control measures are generally only temporarily effective; the area is often re-invaded during the summer, when young moles are on the move from their mothers' burrow system.

To-day moles are usually controlled by poison. The most effective is strychnine, only issued against a permit issued by the Ministry of Agriculture, Fisheries and Food. Moles are the only "pest" against which strychnine is allowed to be used. The usual technique is for worms, preferably *Lumbricus terrestris*, which is the worm most frequently eaten, to be covered with a strychnine-containing powder, and then to be placed in the burrows. It is possible to treat several acres of mole-ridden ground in a short time, and one such treatment is usually successful in killing all the moles. This is therefore cheaper than trapping, for traps have to be set one day, examined the next, and often it takes several days to trap out a colony. If moles must be killed, strychnine poisoning is an efficient method, and if the bait is carefully laid no other mammals or birds are likely to suffer.

Unfortunately strychnine, issued to kill moles, is used for other and less desirable purposes. In most parts of England permits are responsibly issued, and the quantities of the poison made available, though sufficient to kill far more moles than exist in the areas concerned, are not grossly excessive. In Wales and some parts of western England the situation is different. Strychnine issued in Wales every year is probably sufficient to poison the whole human population of the Principality; it would wipe out the moles ten times over, and this is an area where moles are not really serious pests. There is little doubt that much of the strychnine is used for illegal purposes. Hill foxes are considered as pests, and are difficult to control. Some farmers have been found taking a sheep's afterbirth, or a dead lamb, and lacing this with strychnine. The poisoned bait is then hung in a bush or left in some suitable position. Not only foxes are killed, dogs (which wander on to sheep farms, and may worry sheep), cats and possibly kites may be killed. Strychnine abused in this way is a serious menace to man and to wild life. In my opinion its use even for moles should be forbidden.

No one has any doubt that **rats** are pests. The black rat is now confined to towns and ports, the brown rat is an important rural pest. Rats and mice are common on most farms, and damage a great deal of stored grain. They live in open fields during the summer, and return to the farm buildings and ricks in winter. Rat control by trapping, harrying with dogs and even shooting is often practised, but has little permanent effect on the population. Poisoning is the only efficient means of control. In the past many general poisons, arsenic, zinc phosphide and the like have been used. These are obviously dangerous to man, domestic and wild animals. Accidents have often occurred, though it is unlikely that serious effects on wild life were common. To-day the usual rat poison is warfarin, which acts by upsetting the blood clotting mechanism so that the rats bleed, internally, to death. It is thought that this is relatively painless; rats do not, anyhow, provoke the same emotions as rabbits, and moribund animals are seldom a prominent feature of the landscape as are dying and diseased rabbits. Warfarin is particularly effective against rats and mice. They need to take successive doses over a number of days; other mammals, such as dogs, which could be harmed if they also took large amounts over several days, may take an occasional dose of warfarin without being harmed, but the baits are not generally attractive except to rodents. Thus although warfarin is not truly a selective poison, by careful placing of the feeding points, and choice of the material in which it is fed, it acts relatively selectively. It is usually most efficient, and little risk to wild life. The corpse of a rat or mouse killed with warfarin will not harm another animal which eats it.

Unfortunately we have recently seen the appearance, in Wales and Scotland, of rats which are resistant to warfarin. This seems to have been due to a true mutation, not to selection, which is the means of producing insects resistant to insecticides (p. 130). A search is now being made for new poisons. Recently a remarkable substance, norbormide, has been discovered. It is very toxic to rats, which die soon after eating one dose. They are described as having "died in their tracks"; they were not contorted as are animals poisoned with strychnine and the theory is that death is painless. The unusual thing about norbormide is that although it is so effective against rats, it is relatively non-poisonous against almost every other animal. Even mice seem unharmed, and birds have eaten considerable amounts and survived. If norbormide lives up to its early promise, it will be the perfect pesticide against rats. It will also encourage the search for equally selective poisons for other pests.

Fluoracetamide, which achieved such notoriety when at Smarden some infiltrated from a dump and killed cows which drank contaminated water, and which caused the death of dogs in Wales when they ate meat from a poisoned pony, is still used, under strict supervision, for rat control in sewers. Under these conditions fluoracetamide is comparatively safe. The poisoned rats could kill animals which ate their corpses, but there is little risk of this in sewers, although some of the animals may crawl out into the open to die, and may then endanger other carnivorous animals. It is to be hoped that this use of fluoracetamide will not be found essential, and that less dangerous poisons, like norbormide, will soon replace it.

It has been shown that the chlorinated hydrocarbon insecticides are also poisonous to vertebrates, and they have been used against rats and mice. Endrin in particular will kill mice very efficiently. It has been tried in paint; mice ran over the surface and absorbed the endrin. This was not very effective, but when a powder containing endrin was dusted in the mice runs, this proved most successful. It is not normally used in food, so only animals entering the runs are at risk. Those which are too large to enter runs, or which, like birds, seldom do so, suffer little risk. As endrin is such a toxic substance, there are dangers in these methods even when all precautions are observed, and their wide-spread adoption could have unfortunate results. Endrin is very poisonous to man, so some degree of control, which should mini-mise the danger to wild life also, is essential.

Mink have escaped from fur farms in Britain, and are now established in many different areas.   They are particularly numerous in the Dartmoor region, in parts of Wales, in Lanca-shire and some parts of Scotland. It is feared that they may do a lot of damage to poultry, fish, game and to other forms of wild life. It is difficult to be sure whether mink is in fact a serious pest, or whether this is just a victim of zenophobia. We try to preserve our stoats, weasels, pinemartins and wildcats, all of which have similar feeding habits. It may nevertheless be true that the introduced mink is less prone than our indigenous predators to live on rats and mice which are themselves pests. It is said to be a ruthless killer, and mink are accused of killing up to fifty hens in one night.   Like man, and unlike most other predators, it seems to kill for fun. There seems little chance of effective control of the mink by means at present in use (shooting and trapping) so the population is likely to grow. It is important that the effects on other wild life of this new member of our fauna should be studied.

Squirrels are animals which have a popular appeal, and yet are, under certain circumstances, quite serious pests.   Many people accept our native red squirrel, which can do serious damage in young coniferous trees, as desirable, and yet think of the introduced grey squirrel as an objectionable "tree rat." The grey squirrel has indeed spread over large areas of Britain since its introduction in the nineteenth century. It does a good deal of harm in gardens and to trees, though the red squirrel is potentially the greater pest. The range of the red squirrel has decreased since the grey squirrel was introduced. This is often thought of as a cause and effect, but it is possible that changes in forestry and farming are as much to blame for the reduced range of the red squirrel.

For some years an attempt was made to control the grey squirrel by paying a bounty to those who killed the animals and produced their tails for inspection. This cost the Government a lot of money; there is no evidence that the squirrel population was significantly reduced, and the bounty scheme has been ended. Some attempt is still being made to keep down numbers by shooting and destroying dreys. Work is going on to see whether poisons would be more effective. Our anomalous legal system makes this work difficult. It is illegal to poison squirrels in England, but not in Scotland. All factors except the law would make this work more effective in England, where higher squirrel populations are found. Warfarin in a suitable bait, attractive to squirrels and not to most other animals, seems a promising method. If squirrels become more serious pests in England it is likely that the law will be changed so that they can be effectively controlled. It is believed that, properly used, warfarin in a squirrel bait will do little harm to other species.

The **coypu,** a large rat-like fur-bearing animal from South America, was another escape from fur farms. These animals are sometimes exhibited at fairs and inaccurately described as "giant sewer rats." They soon established themselves in East Anglia. At first they were welcomed, as they lived on the vegetation and helped to keep the waterways clear. They increased in numbers, however, and then started to damage crops and destroy river banks by burrowing. The same law that protects the English grey squirrel prevents the use of poison. A systematic trapping programme has eliminated the coypu from most of East Anglia; this is one of the few occasions when trapping has been effective, and has shown that it must be run almost like a military operation to succeed. We have yet to see whether, even with the best planning and adequate resources, trapping can completely

13 *Above*, a sparrowhawk found dead, poisoned by organochlorine insecticides, derived from other birds on which it fed. These had picked up the poison from seed corn dressed with insecticides. *Below*, biological control: a rabbit dead from myxomatosis

14 Pigeon damage and control. *Above*, nests and droppings foul buildings and encourage noxious and dangerous insect pests. *Below*, making Buckingham Palace 'pigeon-proof', by putting strips of a non-poisonous plastic jelly on the ledges. The birds do not like standing on the jelly — so they fly off to perch elsewhere

eliminate this pest. The traps, properly designed and carefully placed, seldom caught other animals. As they were "live traps," had other species been caught they could have been released.

Among the birds, the **woodpigeon** (*Columba palumbus*) is probably our most serious pest. Here again, not everyone objects to this handsome bird, which provides a day's shooting for many countrymen. It is most disliked by farmers growing sprouts and other brassicae, for a flock of pigeons can ruin a field of these crops in a few hours. They also descend on private gardens; many people have found their spring cabbage reduced to skeletons of midribs. They take a substantial amount of the clover intended for winter grazing, though this loss is less easily recognised. They damage newly sown corn by digging up and eating the grain. It has been calculated that they cost British agriculture about £2,000,000 a year. This, though a substantial sum, is small compared with the losses formerly caused by rabbits (£50,000,000) or those still caused by eelworms or insect pests, but individual growers can be ruined by the almost total loss of their crops, so the problem is still serious. We have about 5,000,000 woodpigeons in Britain; there are more blackbirds, chaffinches, starlings and robins, but the heavier pigeon (about 1 lb. in weight) provides much the greatest biomass of any species.

Until recently, the officially sponsored method of pigeon control was shooting. The Ministry of Agriculture subsidised the price of cartridges for this purpose, and the birds were shot by individuals and at organised battus. These were sometimes arranged on a county-wide basis, as it was believed that this would keep the whole pigeon population on the move and prevent escape into areas which were not shot over. We now know that shooting has little value in population control, and it is no longer subsidised. It gave a great deal of entertainment to country dwellers, supplied cheap birds for the pot or the market, but made little difference to the pigeon population. Shooting *could* be effective but only if its intensity were increased at least threefold throughout the winter months. The cost of this would exceed the saving to agriculture. Careful studies have shown that there is always a high mortality during the winter, and that the number of birds which survives is closely related to the food supply. Clover leys are probably the most important source of food, for they provide food at a time when the ample autumn supplies of grain on the stubble, acorns, wild fruits and berries are exhausted. Shooting simply allows another to obtain enough food and survive. For the individual farmer it may not be completely useless. If a flock of pigeons descending on a field of brussels sprouts is met with a hail of shot, it may move on to another farmer's crop; it may only fly over to the other side of the field. Hungry pigeons are hard to deter. Scarecrows, even of the most sophisticated type, seem of little permanent value; they may keep birds away for a day or two but soon lose their effect.

Unlike most birds, which rear their young in spring and early summer, pigeons have been found laying throughout the year. Their peak period for reproduction is August and September, when most food is available. Pigeons almost always nest in trees and bushes. If all the nests were poked out and destroyed in late summer and early autumn, there would be really significant reduction in numbers. Nest poking is now officially encouraged; so far it does not seem to have been thorough enough to be any more effective than shooting.

Many farmers would like to poison pigeons, and they welcomed the deaths from dieldrin poisoning that occurred between 1956 and 1961 (p. 137). Deliberate poisoning using any treated bait is illegal, because of the danger to wild life, and to man from eating poisoned corpses. However, work is now in progress on the use of bait treated with narcotics. The birds eat this, and fall down stupefied. Tests have shown that very large numbers of pigeons can be captured. Few other species are affected, and almost all of these will recover and fly away apparently unharmed. If large seeds, like beans, are used as bait, few other birds are able to swallow them, so the treatment becomes comparatively "selective." With skilled operators, with time to collect narcotised pigeons and release other birds, this method may help to reduce pigeon numbers, but control will probably need the co-ordination of nest destruction, shooting, narcotisation and some control of crops which provide winter food. The wood-pigeon is only too well adapted to the environment that is produced by modern British agriculture. Incidentally, though it is true that sparrowhawks and peregrines kill many pigeons, there is little evidence to support the view that the decrease in the numbers of these predators has been an important factor in allowing pigeon numbers to reach pest proportions. Pigeons in regions like East Anglia are not in fact much more numerous in 1966 than they were twenty years earlier when predatory birds were more numerous.

Visitors to London admire, and feed, the pigeons in Trafalgar Square. Similar populations are found in many cities. These are not, as is often thought, woodpigeons which have invaded the cities, but the descendants of escaped domestic pigeons which are really rock-doves (*Columba livia*) and a quite different species. These **feral pigeons** are a very motley crew. They have inter-bred with domestic flocks, and have been frequently reinforced with further escapes. They show traces of all the colours and forms known to pigeon fanciers. When they get too numerous they foul the buildings, and may do considerable damage to the structure. They also invade urban gardens. The owners of most buildings would prefer them to nest and roost else-where.

Public sentiment makes control by shooting or trapping impossible. These birds could in fact be easily controlled by these means though they are so ineffective against woodpigeons. In some cases narcotising baits have been successfully used (e.g. among the unsentimental French), with negligible risk to other species except perhaps house sparrows, which may themselves have the status of pests. If feral pigeons were not deliberately fed, they would not breed so successfully and would probably not reach numbers sufficient to warrant control. Attempts to drive the birds from city centres by noises from tape recorders, stuffed owls and cats have all been tried with little success. Many buildings have now been made almost completely "bird-proof" by the use of strips of a plastic material applied on to the ledges where the pigeons usually roost. This does not harm or entangle the birds; the yielding, jelly-like material makes them uncomfortably insecure, and they soon depart from a building which has been adequately treated. This method does nothing immediately to eradicate the pest, but if pigeons are unable to roost in the centre of the city they will probably come less often in search of food, and in time breeding will be less frequent and the population will decrease. **Starlings**, which also roost in huge numbers in some urban sites, can also be got rid of using the same material. This technique is one which controls pests by making part of the habitat inhospitable. It does not affect other species, nor does it contaminate the environment. It seems, therefore, that this plastic strip is also a "perfect" pesticide.

The **house sparrow** is another pest, not only in Europe where it is indigenous, but in America and Australia where it has been introduced. House sparrows are comparatively tame, and enter buildings after food. They nest in the girders in stores and even in bakehouses, when they bring dirt into the buildings as well as damaging food. The trouble is often increased by kind people feeding the birds. Sparrows are also pests on farms. They consume a lot of grain in stores, and, shortly before harvest, vast numbers may invade the fields and strip plants of corn as it ripens. The amount taken is substantial; on small experimental plots at Research Stations the loss may be sufficient to render useless experiments which have taken a great deal of the efforts of scientists. Shooting (with special small shot) can have a little effect. Crops can be netted, but this is usually too expensive. Various types of bird-scaring devices, including automatic "bangers" have some use, but in general it has proved impossible to protect ordinary commercial crops at an economic price.

Control on farms still presents many problems, but sparrows have been almost eliminated from some bakeries and similar places by the use of narcotics. A suitable bait is laid early in the morning, usually at a week-end when the building is not in use and when it is easier to eliminate other sources of food. A few hours later the sleeping sparrows are collected and killed humanely. A few members of other species may be discovered – in practice less than half of one per cent of the total – and these almost always recover and can be released. This treatment can only be applied by skilled staff under licence, so harm to wild life is reduced to a minimum.

A number of other wild birds, including **carrion crows** and **gulls**, are thought of as pests in different parts of the country. Here again we come up against the problem of what is a pest and what is wild life. Any wholesale measures against these species would probably affect many others. The most promising method against gulls, which have certainly increased in numbers in recent years, is to kill eggs by dipping them in a poisonous liquid (e.g. an oil+formalin). The birds continue to incubate the eggs, and may not breed for a whole season. If the eggs are removed many will lay a second, successful, clutch.

Birds on airfields may endanger the aircraft. Attempts have been made to eliminate worms from the grass, as these attract many birds. Unless very high doses of poison are used over a wide area this method has little success. The organo-chlorine insecticide chlordane is used to kill the worms; this method does not receive general approval as it may contribute to environmental pollution. Various noises – tape-recorded alarm notes of birds, for instance – have been used with limited success. The most interesting method now being tried is to use trained falcons to drive the offending birds away.

In recent years the **bullfinch** has become an increasingly serious pest in orchards, where it destroys fruit buds in winter. Sometimes the crop may be reduced almost to nothing. We do not know why the bullfinch has become so much commoner. It is one of the comparatively few species of bird which is found in modern orchards, and it is probable that changes in the technique used by fruit farmers have encouraged its successful breeding. There is no evidence here that the disappearance of hawks is a main cause. Seed-baited live traps, sometimes with a decoy bird, catch many, and give some relief and do no harm to other species, as these can be released unharmed. Various inexpensive scaring devices, such as almost invisible threads covering branches, give a fair degree of protection. To many who do not make their living by growing fruit the bullfinch is such an attractive bird that the damage it does is accepted without complaint.

CHAPTER ELEVEN

THE FUTURE – IMPROVEMENT OR DISASTER?

I have shown a number of the ways in which we are polluting Britain. Everyone agrees that this is a bad thing, and would like it to stop. We try to prevent noxious pollutions in many different ways. In theory at least if any person suffers damage to his person or property, he can obtain damages in a court of law from the person who is responsible for the pollution. Unfortunately, though damage is common enough, successful prosecutions are rare. It is generally difficult to prove where the pollution came from, and it is not always easy to define at all precisely the nature of the damage. An exception is when damage is done by herbicidal sprays drifting on to susceptible crops. Here it is often possible to prove who was responsible, and damage from weedkillers can generally be accurately diagnosed. The injury is due to an action which took place at an exact time, and witnesses can sometimes be found who actually saw the spraying taking place in a high wind. When it comes to damage from chronic pollution from smoke, legal redress is unlikely. Even the pollution of a river, with the killing of fish, from industrial effluents may be difficult to prove, as before the scientist called as an expert visits the site the current may have carried the poison far down the river, the dead fish may have been removed, and arguments based on reduced fish populations may not convince the court. In general the law does little to protect the public from damage by pollution.

In specific cases the law can be more effective. There are whole series of Acts of Parliament, some of them hundreds of years old, on the subject. Certain noxious discharges are controlled by the Alkali Acts, and food contamination with poisons, for instance arsenic (used on fruit as an insecticide) is regulated by the Food

183

and Drug Acts. Smoke emission from factories and house is forbidden in specified areas, and in these only smokeless fuel can be used in domestic grates. Radiation is rigorously and successfully controlled by law. The effect of this legislation in the last fifty years has been important; without it our much increased population, and our intense industrialisation, would have had disastrous results. There is nevertheless clearly the need for further action to deal with many continuing abuses. Only an informed public opinion can have any real and permanent effect.

Pesticides provide special problems. Some insecticides and herbicides are acutely poisonous to man, and the Agriculture (Poisonous Substances) Act, 1952, lays down regulations to protect workers in agriculture and horticulture. Protective clothing must be worn when the most poisonous substances are being used. This legislation has been very successful; cases of workers being poisoned are at least a hundred times less common than fatal accidents involving farm machinery.

We try to prevent dangers to the consumers of agricultural produce which has been treated with pesticides, and to wild life which may be endangered from agricultural chemicals, by the Pesticides Safety Precautions Scheme, which is a typical British invention. This is a voluntary scheme, freely entered into by the manufacturers of insecticides, herbicides and fungicides. These firms agree that, before a new chemical is put on the market, they will supply details of its properties, including its toxicity, to the Ministry of Agriculture, Fisheries and Food. These data are considered by the Advisory Committee on Pesticides and Other Toxic Chemicals. This committee relies largely on the advice of its Scientific Sub-Committee, a body of experts including ecologists, but excluding any trade representatives. Since this scheme has been in operation, no new pesticide has been marketed which has been shown to have seriously harmful effects on wild life. The system may not be perfect, but it does seem to work, though it places an increasingly heavy burden on our scientists.

Pollution control has done a lot to restrict damage to human and other forms of life. We need, however, still to assess the present problem. We may be making too much fuss about some process which in fact is relatively harmless, and too little about another which is doing real damage. But our eventual objective must be to try to put a stop to all pollution, for it all does *some* damage.

Air pollution is a serious problem in all industrial countries. It is said to cost the community in Britain some £250,000,000 annually, though I am not sure exactly how this figure was calculated. It is satisfactory to know that, at long last, we are taking some effective action to reduce air pollution in our cities. Notwithstanding the rise in human population, and the great increase in recent years in industrial productivity, the air in our major cities is cleaner than it was fifty years ago. We have set up "smokeless zones," where conditions are much improved, and progress, albeit much too slow, is being made to extend these. Fog and smog are still much too common, but we never get conditions approaching the "pea-souper" of Victorian London. We could easily reduce particulate contamination much further, for it comes mostly from open fires in domestic grates, if we manufactured enough smokeless fuel to make it possible to insist that it alone was used at least in towns. The improvement from this one measure would be large, the cost comparatively small.

Though particulate contamination is falling, and could easily be reduced further, the sulphur dioxide content of the air is rising. Oil-fired furnaces produce this SO_2, and it is difficult, but not impossible, to prevent it entering the atmosphere. Factory chimneys are getting taller and taller, and this means that the gases are discharged at high velocity into the atmosphere. The concentration of SO_2 from a modern factory may be very low in its vicinity, but wind may carry the gas to other places. It is generally believed that at the levels present to-day, SO_2 is harmless to animal life, though sensitive plants may be damaged. It seems to me that air pollution may be one subject about which we should, as a nation, be more worried. If plants sometimes suffer acute damage, then it is quite possible that the long-term effects of low levels of pollution are harmful to man. We know that smog can be lethal to sufferers from bronchial complaints; it is hard to be sure that the rest of us do not suffer some ill effects. As I have already pointed out, it is man rather than animals that is most likely to suffer from man-made atmospheric pollution, so he might be expected to take the matter more seriously. Even if it is costly, we should insist that factories, and even domestic oil-fired burners, reduce their emission of SO_2. There is one hopeful feature. New sources of energy, including atomic and hydro-electric power, do not normally pollute the air.

Motor transport is another source of atmospheric pollution. Although there is much criticism of the black fumes which belch out of many of our lorries, these diesel engines are in fact less likely to cause dangerous pollution than are the apparently cleaner petrol engines. Diesel engines only produce smoke when they are improperly used, and this nuisance could be stopped at once if the existing laws were rigorously enforced. Even now the damage both to man and wild life is probably more apparent than real. It should be remembered that diesels are safely used in the confined tunnels of mines where petrol engines are impossibly dangerous. Motor transport pollutes the atmosphere with carbon monoxide, lead products and a whole range of organic substances. In narrow crowded city streets car drivers stuck in traffic blocks may experience dangerous concentrations of carbon monoxide, though under most other circumstances diffusion is great enough to obviate damage to man or other organisms. Lead may be concentrated in roadside vegetation, though few instances of poisoning from this source have been reported. Nevertheless, lead is an insidious poison, and the chronic effects of low doses are serious, so this hazard needs to be kept under observation. It might be as well to let our engines "knock" if our health is endangered. The various hydro-carbons in car exhaust gases are responsible, under the peculiar climatic conditions combined with the high density of enormous automobiles, for the famous Los Angeles smog; similar effects are seldom if ever observed in Britain, and even with more cars our climate may save us from this trouble.

Various other local pollution problems arise from time to time, such as fluoride from brickworks. These seldom affect either man or wild life except over limited areas. Our existing laws could prevent damage from these causes.

The one form of atmospheric pollution which has been shown to have serious effects on man is the self-induced pollution of the lungs by cigarette smoke. Fortunately this does not affect other species, which do not indulge in our vices. Smoke is a poor repellent for midges and mosquitoes, which seldom receive a lethal dose! Incidentally it has been suggested that some other pollutants, common in cities, act as an "urban factor" and increase the danger from cigarettes to town dwellers compared with the risk to those living in rural surroundings. Recent work tends to minimise the importance of this alleged urban factor. But if there is indeed an urban factor, its cause needs to be isolated, and its effects on all species evaluated.

There is no doubt that many of our rivers are much less foul than they were a hundred years ago, even though they may consist of little else than treated effluent from sewage works. This improvement may continue, but it will probably not give us back rivers with the fish and other wild life found in clean and natural waters unless our whole water policy is changed. Also as the demand for water increases, a little for human use, a great deal for industry, if this is taken from the clean headwaters there will be little to flow down the natural river beds at many times of the year. A quite new approach is needed.

When a cheap and efficient method of producing fresh from sea water is perfected, it will solve the problem. Let us hope that this will be soon. In the meantime we must reconsider our needs. "Safe" water can be produced from the lower reaches of our rivers. This is certainly fit for industrial purposes, and for most domestic ones as well. We only drink a few pints a day; we use hundreds of gallons to wash cars and flush water closets. Perhaps we should have our drinking water delivered in bottles, like our milk. Already the whisky connoisseur imports special water to dilute his beverage, if he cannot count on the pure water from Loch Katrine coming out of his tap.

Whatever the source of our water, our waste, from homes and

factories, has to be got rid of. At present it goes into our rivers, hygienically treated no doubt, but still containing many noxious substances as is demonstrated by the creation of detergent "swans." Progress is being made in more efficient purification, but I am sure that the only satisfactory solution will be to devise other methods of disposal. This has had to be done for radio-active waste. We could turn our organic waste into fertiliser, which would be an economy as well as a reduction of pollution. Factories may construct waste pipes which run far out to sea. We may devise what is in fact a "sewer grid" for objectionable wastes. At present we could not do this for all waste water, as we would then have no rivers running most of the year. Only selected waste could therefore use the grid. If, however, we pump desalinated sea-water in a water grid, all the waste can go into the sewer grid and our rivers can revert to their pristine condition!

Pure rivers will cost a great deal of money, but it is in our power to have them. It all depends on our scale of priorities. We must always remember that pollution of water has a far greater ecological effect than the same degree of contamination in other parts of our environment.

Though air and water pollution does a lot of damage it is unlikely to kill many of us, and much of our wild life will probably survive. There is a distinct chance that we will all be killed by radiation; one madman at the controls could start a nuclear war and wipe out human, and most other life all over the world. But though radiation is, potentially, the greatest risk, it is probably the least serious source of pollution to-day. It is one sort of pollution we have, so far, been able to control. We have been prepared to spend vast sums on this control, and we should continue to do so. As nuclear energy, for legitimate peaceful pur-poses, increases in importance, there will be greater chances of accidents, with the release of radio-active pollution. Even present low levels have probably had harmful effects on a tiny minority of human beings; a tiny minority, but many thousand indi-viduals. It is difficult to know whether wild life has been affected, except near to nuclear explosions. The only sensible course is to do everything possible to reduce future contamination to the minimum, and to hope that no major catastrophe will occur.

So far we have considered pollutions that are the by-products of civilisation. They must be controlled, and controlled stringently, as the world becomes more crowded, if life is to be worth living, and if we are to share the globe with other forms of life. We must now consider how we can mitigate the effects of our deliberate spreading, with the best of intentions, of poisonous pesticides throughout the country.

Herbicides, fungicides, insecticides and other toxic pesticides are essential to British agriculture to-day, and without them our present farming pattern would disappear. It is impossible to say just how much damage would be done if we had no pesticides at our disposal. It is estimated that, even with an annual expenditure of £15,000,000 on these substances, we lose about eighteen per cent of the possible yield of our farms. This loss amounts to as much as £140,000,000 from crops and £150,000,000 from animal husbandry. Without pesticides some crops, e.g. field beans, would probably never be grown. The annual loss would certainly be over five hundred million pounds if pesticides were not used, unless far-reaching changes, which might themselves reduce yields, were made in agricultural practices. Incidentally it may be interesting at this time to mention that we spend in Britain each year under ten million pounds on agricultural research, and under one million on wild life conservation.

The disappearance of all pesticides would profoundly affect British agriculture, but substantial crops of various kinds could still be produced. In many parts of the world, including North America, and the tropics, the situation is different. There it is common for pests and disease to render a crop a total loss; we are almost always able to salvage something. It is not surprising, therefore, that we have escaped some of the worst side effects of the use, and abuse, of dangerous pesticides.

We have seen that most of the herbicides in use to-day are relatively non-toxic to invertebrate and vertebrate animals. Birds and mammals can seldom be harmed, directly, by their use. Even after regular use over several years on the same patch of ground, weedkillers do not seem to harm the soil fauna. Changes do occur, but these are usually less than those caused by other agricultural processes like normal cultivation. Interesting changes in the microflora have been reported. Herbicides like MCPA may be more rapidly broken down in soil frequently treated than where the herbicide is used for the first time, as the bacteria tend to "learn" the process. This is likely to reduce rather than increase dangers of pollution. The selection of herbicide-resistant strains of weeds is an important possibility, though there is little evidence of this happening as yet in Britain. Such weeds could affect not only arable fields but also the surrounding country.

The importance of weedkillers to wild life is that they may profoundly change the habitat. MCPA and similar substances used on arable crops have minimal effects, provided care is taken to avoid damage from spray drift. As these herbicides have only replaced hand weeding, their use is generally accepted. The real danger of herbicides is that they are so efficient. It has become so much easier to clean up patches of rough ground and scrub, which may now be important wild life sanctuaries. With labour costs too high to continue hand-cutting of roadside verges, cheaper measures are being introduced. Here weedkillers may not always be the worst solution, but much more study is needed. The use of Paraquat to improve rough grazing is a typical case of a relatively harmless substance which can be used to do great damage. Here, as in so many other situations, we have to decide just what we want to preserve. In the past this has been left to chance. We even have one government department paying a subsidy to a farmer to spray and cultivate an area which another department is trying to preserve in its original state. Without efficient herbicides the risk of damage is very much less, so some sort of control is needed even of substances which are not, technically, poisons, if wild life preservation is to succeed. The direct danger to man of modern herbicides would seem to be minimal.

Fungicides are widely used in agriculture and horticulture. Almost all seed corn is treated with organo-mercurials. So far British experience suggests that these are harmless to wild life, though Swedish observations make it desirable to re-examine the situation. Other fungicides are used in more restricted situations, and we have little reason to suspect that they are dangerous.

Insecticides are the substances which have caused most concern in Britain, and throughout the world. These are all, by definition, poisonous to insects, and they are, to a lesser or greater degree, poisonous to other animals also. In the past there have been unpleasant incidents clearly implicating the more toxic organo-phosphorus insecticides, but the substances in this group now in common use are much less poisonous to vertebrates and probably do them little harm. They are easily decomposed, and so do not yield long persisting toxic residues. The search for more selective organo-phosphorus compounds continues, with some success, as does also the discovery of insecticides with some persistence (perhaps for a season) but which do not cause environmental pollution which lasts for years.

There is one thing which may delay the discovery of really selective insecticides. Such substances, effective against a single pest and otherwise harmless, may be discovered. They will, by their very desirable qualities, have a very limited market, while a "broad spectrum" insecticide useful against many pests (and lethal to many animals which are not pests) will sell in far greater quantities. Most of our insecticide research has been done by industry. It may cost a million pounds to develop a new compound to the stage when it is marketed; a great deal must be sold to recover this cost. It is likely that in future very much greater sums of government money will have to be spent on pesticide research if more selective, and less profitable, substances are to be made available.

It is the persistent organo-chlorines that are the villains of the piece. There are well authenticated reports of serious incidents affecting vast numbers of birds in America and other parts of the world, where DDT, dieldrin and several other substances have been implicated. Not only have many individual birds been killed, but also various species have been eliminated from large areas and some are in serious danger of extinction. In Britain we have, on the whole, been more fortunate. Until we stopped using dieldrin-dressed seed in spring, great numbers of birds were poisoned and died, but since 1961 this cause of death has been very greatly reduced. We have good evidence to show that several predatory birds have been seriously affected. Other than this, there is little incontrovertible evidence that our wild life has been seriously damaged. Under the conditions which exist in Britain most of the insecticides which have caused damage elsewhere have probably been relatively safely used. I do not in fact believe that DDT or BHC have yet had any serious effect on British wild life. Dieldrin, aldrin and heptachlor on the other hand have proved dangerous, and the recommendation that their use be restricted was justified. I personally believe that they should be removed from use as soon as substitutes as effective, but less persistent and less toxic to wild life, can be found. I have come to this conclusion reluctantly, for I realise that these are substances of great value to farmers, but they are just *too* persistent. They are still widely used in many under-developed countries, and I have no doubt that they have again proved their worth. Even here, however, they should be replaced as soon as possible, for the persistent residues must be building up at a level which is potentially dangerous.

The main worry caused by the persistent organo-chlorine insecticides is that they have polluted not only all of Britain, but the whole of the world. It is difficult if not impossible to find a field in England which does not contain a detectable amount of pesticide, or a bird or mammal without residues in its tissues. There is no doubt that many have greatly exaggerated the danger of this pollution. Most of the residues are so low that their discovery is a credit to the chemist rather than a menace to wild life. Nevertheless, residues at a toxic level *are* found, commonly in raptorial birds, not infrequently in other species. The general level of contamination is high enough to warrant concern, particularly as concentration of a persistent chemical is always possible. For radiation, a lower level of general pollution for the population is accepted than is tolerated for the few individuals who are at risk from their employment in the industry. For pesticides, we should similarly be concerned about any general increase in pollution even if this does not necessarily damage particular individuals. Ecologists in particular are concerned about this general pollution. They point out that effects in the field cannot be fully foretold by a study of toxicological data. This is of course true, but ecology is still such an inexact science that none of us can be dogmatic, and the need for more research is constantly being revealed.

Some writers have suggested that low levels of pollution, particularly with pesticides, may endanger the health of the individual by causing cancer and the health of the species by affecting its genetical make-up. It is true that some pesticides are related, chemically, to known carcinogens, but there is no evidence that any widely used pesticide, applied in the proper way, ever causes cancer. Nevertheless, this risk must always be borne in mind. The dangers of long-term exposure to apparently harmless radiation levels proved far greater than had been expected by the experts, and on an analogy with this we will not know until too late whether exposure to a few parts per million for twenty years of DDE or other apparently harmless chemical has had some damaging effect.

The genetical make-up of animals and plants could be affected by chemical factors in the environment, including pollutions from industry, and pesticides of all kinds, in two main ways. First, mutations could be induced. We know that mutations occur with increased frequency when organisms are exposed to radiation (see p. 63) and to a variety of chemical substances. Thus colchicine has been shown to cause polyploidy in plants, and this may produce larger and more vigorous organisms. Other substances are described as "radiomimetic," and, like radiation, affect the chromosomes and cause gene mutations. Laboratory experiments have shown that some pesticides (e.g. BHC) can produce polyploidy, and effects on mitosis, particularly from high concentrations, have been reported with several insecticides and herbicides. We know that the majority of gene mutations are deleterious, and in a wild population they are normally eliminated. In man, harmful mutations may be kept alive, and

P.P.   N for this reason the risks of their arising from man-made radiation have been the cause of the strict regulations governing the disposal of radio-active wastes. There is at present no firm evidence that pollutions of the kind and at the level we have been considering have produced mutations in man or in wild life. However, the subject is one which requires to be constantly studied, and new chemicals should if possible be tested as mutagens as well as poisons. It is dangerous to be too dogmatic, for there is always the possibility that mutations may be produced more readily in one species than in another, and that the most susceptible have not been included in any screening tests.

One point regarding mutations in animals or plants which are economic pests should be noted. Some people have expressed the view that chemically induced mutations will be likely to be resistant to the toxicity of that particular chemical, and this has been given as the reason for insecticide-resistant strains of pests. It is not impossible that resistant forms should appear under these circumstances, but there is no reason to believe that in general a mutation is related to its cause in this way. Thus a housefly resistant to DDT might arise as a mutation spontaneously, or be induced by radiation, and there is no reason to blame its initial appearance on the chemical, though its persistence would be due to the effects of the chemical on the remainder of the population. However, most insecticide-resistant strains seem to have arisen without the need for specific mutations.

If, and it is still *if*, any pollution from pesticides or other pollution has a mutagenic effect, it is man who is most likely to be harmed. The next group of organisms at risk is that of pests themselves, including weeds, insects and fungal pathogens, as they will experience the most prolonged exposure to the highest concentrations. Mutations here would most likely be rapidly eliminated; the chance of producing "super-pests" of even greater danger is remote, except in the case of polyploid weeds which might be more difficult to control than normal ones. So far the appearance of dangerous polyploid weeds has not been attributable to chemicals in the environment, but this risk must be borne in mind. Finally, among wild life chemically induced mutations are not very likely to occur, as exposures will not usually be high, but if they occur they will probably soon be eliminated by natural selection, though it is possible that in a rare species where the total numbers are small a harmful mutation could weaken a substantial part of the population and so hasten its total extinction.

The second way in which pollutions do in fact produce hereditable effects is by producing conditions favourable to certain already existing varieties of a species. The selection of melanic moths in industrial areas (see p. 41) is one familiar instance, and the production of insecticide-resistant pests is another. These changes can be very important, but they are similar to those caused by any environmental change, whether produced by man or by natural events. Except that pesticides defeat themselves when they are responsible for the emergence of resistant strains, there is nothing more sinister in this than there is in any man-made environmental change.

The best way to reduce possible environmental damage from pesticides is to reduce their use to a minimum. In this connection, great importance is placed on the use of "Biological Control." This term covers a great many processes. To some it means primarily the use of beneficial insects, predators and parasites which keep pest numbers down and so reduce or prevent damage. It also covers the use of insect diseases, of various techniques to induce sterility or upset sexual behaviour and cultural methods which make crops unattractive. Emphasis is also put on "integrated control"; this means different things to different scientists, but in general implies a combination of chemical and biological methods, with, if possible, the minimum use of toxic pesticides.

Biological control has had its spectacular successes: these have so often been described that I need only mention the subject briefly. The citrus scale insect, introduced from Australia to California, was ruining the industry. An Australian ladybird beetle, which fed on the scale insect, was transported to California and gave complete control. This was effective until insecticide sprays (for other pests, not controlled by the ladybird) killed so many ladybirds that the scale again achieved pest numbers. The prickly pear, introduced to Australia from America, was a troublesome weed in pasture. A caterpillar from America wiped it out almost completely. These two instances are typical examples. A pest, or a weed, is introduced into a new habitat without bringing the insects which keep it in check in its home territory. The pest gets out of hand, the balance is restored by a suitable introduction. The great advantage of this method is that control may continue, without any help from man, or cost to the farmer, almost indefinitely once the useful insect is established.

There have been numbers of other instances of total success. There have also been many cases where a parasitic or predatory insect has made considerable inroads into pest numbers, and given substantial results without giving complete control. Some critics of biological control have, unfairly, treated such instances as failures. There have, of course, been many failures, but comparatively little money has been spent on this work, and the dividends from the successes have been enormous.

Britain does not appear a promising place to produce such dramatic results. First we have few really devastating insect pests, and secondly our pests, such as they are, are mostly indigenous. They are not more serious because they are already subject to considerable biological control; as we have seen already, excessive use of insecticides can make a pest (red spider mite, cabbage root fly) more serious, by removing the beneficial insects which were controlling it. Biological control is most likely to succeed in a perennial crop, such as apples, or where an annual crop is grown year after year in the same situation. Many of our worst pests in annual crops grown in rotations are bad targets for parasites and predators, but work on this subject should be intensified. Insects introduced from other countries may control our pests even better than do indigenous parasites and predators, and may reduce the need for much chemical control, even if the chance of this eliminating insecticides in all British farms is remote.

It is often denied that chemical control can ever produce a real solution, so we should remember cases when it does. The

Colorado potato beetle, introduced into Europe from America, does much harm on the Continent, but although specimens have often reached England, and have temporarily established infestations, these have always been eradicated, usually with DDT. It would be difficult to find a predator which could, in the same time, prove so efficient. In general, however, it is true that pesticides do not permanently solve a problem, and their over-use may eventually make things worse. We would all like to be able to discover an ecological solution, including some way of growing our crops without the need for a constantly changing spectrum of pesticides. Progress in this direction may be possible, but as our whole agricultural pattern changes, so will the problems due to pests and those arising from different types of cultivation. Some types of pesticide will, in my opinion, always be needed, and we are right to try to ensure that these are as harmless as possible.

Virus and bacterial diseases occur among insects, and have been used to control outbreaks of pests. Some are relatively specific, only affecting one insect, so beneficial forms are un-harmed. Some progress has been made to control pests in Britain in this way; fortunately a diseased cabbage white caterpillar does not arouse the same emotions as a rabbit with myxomatosis, which, I again repeat, was the most spectacularly effective case of biological control. Some bacteria produce crystals which are virtually insecticides, and should really be considered as such, though their use is often quoted as biological control. These crystals have the virtue of being non-poisonous to vertebrates, and of being relatively specific in their action against insects. All these methods, which could reduce the use of chemicals, obviously need to be studied further.

One method of pest control which captured the public imagination is the "sterile male" technique. In several parts of America the screw worm fly, whose larva attacks the living flesh of mammals, has been controlled by this means. Vast numbers of flies are bred in captivity, and are then subjected to doses of radiation sufficient to induce sterility without loss of virility. The males pair with the females; this process is accomplished only once in the insect's life, so sterile males cause the production of infertile eggs. Many of us thought that this ingenious method would not work in practice, but it has done so for a few pests. Attempts to use the method in Britain have so far failed, and entomologists have not found a suitable pest with habits which could make control likely. Work is also going on to find chemicals which induce sterility. I do not favour this approach. Such chemicals are really just chemical insecticides. They have to be applied to the habitat like poisons, and I think that they are more likely to have unpleasant side-effects, including poisoning and reproductive upsets in man, than are most ordinary insecticides. Used selectively in bait they may be less dangerous.

Some female insects produce "pheremones," substances which lure males from long distances. The Gypsy moth, a pest of forest trees, is such an insect. A synthetic chemical with similar properties has been produced, and successfully used to capture huge numbers of male Gypsy moths. These synthetic pheremones will not pollute the environment, and are specific. Unfortunately most pests cannot be caught in this way, as they do not behave like Gypsy moths. Another approach has been to produce chemicals which resemble those which cause the sexes to meet, even in insects which do not respond over long distances. If an area is saturated with such a chemical, it is assumed that insects will be confused, and successful mating will be prevented. Some promising work on these lines has been done, but it is doubtful whether these techniques are really safer than non-toxic non-persistent insecticides.

Insectivorous birds eat pest insects, and this must reduce the population. It is difficult to discover whether this is often a significant method of control. It is often suggested that insecticides have killed birds which previously controlled insect pests which are now much more serious. There is, to my knowledge, no proof of this, but it may well occur. I have known birds remove all the cabbage white caterpillars from plants when I have needed the infestation for experimental purposes. Some birds which eat pests at one time of year, and so are "beneficial," attack our crops at other times, and so then become themselves "pests." In many cases, including our common rook, it is difficult to know which role is the most important.

Scientists are trying to devise many other non-chemical methods for pest control. I have already mentioned the use of heat to disinfest lousy garments and grain with mites or beetles in it. Heat has also been successfully used to treat timber; unfortunately it generally remains susceptible to reinfestation, and as wood has such a long life, this is more serious than with foods which have only to be kept insect free until eaten. Radiations of various kinds are being tried. These can be dangerous to operators, they may leave the treated substance radio-active, and they may taint it. Recently a different type of radiation has been tried. A sort of "death ray" has been used against stored products and timber pests. The grain or other substance is exposed to radio-frequency electric fields. This kills the insects, because they absorb the energy differentially from the medium in which they are living and are in effect killed by the heat produced. So far this method has proved more expensive than fumigation. It has limits; in timber other objects including small pieces of metal, perhaps splinters from saws or bits of nails buried in tree-trunks, cause even greater heating and the danger of fire. Nevertheless, methods of this kind are likely to be important in the future, and will help to reduce the risk of pollution.

The safest method of pest control is to grow plants which are not attacked or which can tolerate attacks without suffering economic damage. Our plant breeders have had some success. We have varieties of sugar beet which are not attractive to aphids, we have strains which are tolerant of virus disease. Much more research on this subject, which could make it quite unnecessary to use chemical pesticides for the crops concerned, is obviously needed.

Farmers and gardeners know that they can avoid or reduce insect attack and damage by planting crops at the right time. Broad beans planted early develop their crop before the aphids do any damage. Oats sown in February are too big to be attractive to the frit fly when these insects are egg-laying in May. Early sown winter wheat is strong enough to survive the wheat-

bulb fly larva's attack when it emerges from its eggs in March.
All these cultural methods are important and reduce the need
for insecticides. Recently it appears that we may learn to exploit
the effects of all kinds of cultural techniques on the relations
between pests and their hosts. There are people in Britain and
other countries who follow what is often called the "Organic
Method" of farming. They believe that chemical ("artificial")
fertilisers have serious faults not shared with compost and other
forms of "natural" manure. In my opinion too much has often
been claimed for the results, but I am convinced that the subject
needs a thorough and scientific investigation, and that many
orthodox scientists who are scornful of this work are just as
unscientific as their opponents. I have seen too many instances
where plants grown organically have had lower infestations of
pests than apparently healthy crops nearby grown by other
routines. Man has undoubtedly produced conditions in which
pests flourish, but completely natural conditions do not always
ensure against insect attack. Locust swarms develop in areas far
from human interference, and the migrating swarms can devastate
advanced agriculture or virgin forest. Oak trees in ancient forest
can be defoliated by *Tortrix* caterpillars as seriously as can trees
in intensely managed plantations. But there are undoubtedly
many problems to study in relation to cultivation, management
and pests, and problems whose solution will be important to
farmers and foresters and gardeners alike.

"Integrated control" in which chemicals, beneficial insects and
cultural methods act together and not in opposition is really
common sense, but it can only be effective if we understand a
great deal about the ecology of a pest. Only a few pests have
been studied in the detail needed. If we want clean crops and
profitable agriculture, we must save money (and avoid unnecessary pollution) by only using the minimum amount of insecticide
and only using it when it will do good. Most people are now
opposed to the idea that if there may possibly be an outbreak
then pesticides should be applied as a prophylactic. This
was often done in the past with fertilisers containing aldrin.
These killed wireworms while they added the chemicals pro-

vided for plant growth, but they were often used when there were
few pests and so the soil was unnecessarily contaminated and
beneficial insects were eliminated. If we could always foretell
when an insect would reach pest proportions, and when it would
not, we could reduce insecticide use very considerably.

Private gardens pose particularly important problems to-day.
They are increasingly valuable refuges for wild life. Many birds
are found there in greater numbers than in any other habitats.
A great deal of damage is done by the over-enthusiastic amateur
gardener if he is too liberal in his application of pesticides. The
Ministry of Agriculture produces a shilling booklet called
Chemicals for the Gardener; I find it a rather frightening compilation.
Though excellent advice on the avoidance of dangerous practices
is given, the reader finds himself advised to apply an enormous
range of substances. He should use 2,4-D or MCPA on his lawns,
simazine and similar long-lasting weedkillers on the path, dalapon
to kill couch grass, Paraquat instead of weeding the beds. Fruit
trees may be sprayed with DNOC, tar oil, BHC, DDT, malathion and
a host of other chemicals. Practically every other crop should
receive one or more of these same insecticides. The question is,
is this all necessary? The farmer has to control pests if he is to
make a living. Most of our shops insist that fruit is unblemished
and uniform, even if few questions are asked about its flavour.
But does a little damage matter to the amateur, unless he wants
to exhibit at his local flower show? No doubt if he sprays and
sprays he will have "clean" crops, and a higher level of pesticide
in his, and his neighbours', tissues. He may find his wild birds
are dying, and that eggs are left cold in the nests. Sometimes he
will find that if he forgets to spray again he will have even more
serious damage, for his garden will be devoid of predators. An
unsprayed garden usually gives reasonable crops. A few apples
may contain grubs but these can be cut out – some people think
the rest of the flesh is all the sweeter! A garden, with diversity
of trees, shrubs, grass and crops is unlike a huge area of agricultural monoculture. There are many beneficial insects. Biological
control goes on all the time, not a hundred per cent successful,
but sufficient to satisfy most householders who suffer more often

from a glut than a shortage. They will find their friends appreciate their produce all the more, even if it bears a few marks of insects' jaws, if they know that it is free from pesticide residues. And there are now even some commercial shops which have discerning customers willing to pay more for slightly scabby apples with a good flavour and no organo-chlorine content. If our gardens could be kept free from at least the more dangerous pesticides, the effect on the wild life of Britain might be enormous.

In many countries, particularly in the tropics, insect-borne diseases (malaria, plague, sleeping sickness, etc.) are of major importance. Even in Britain biting insects may be pests. Medical entomologists are usually more wholesale in their use of insecticides than are agricultural entomologists. We believe it is right to try to eradicate an insect which causes disease to man. Crop pests only need to be reduced below levels where they cause economic damage. To-day medical entomologists would mostly prefer not to use insecticides, which give only temporary relief, and would rely on changes of the habitat to make it unsuitable for pests and disease vectors to breed. This can be an admirable solution, but it can also bring conflict with wild life interests. Conservationists are very worried by the disappearance of "wetlands," marshes and swamps famous for bird life. Some areas have been lost to agriculture, some have been drained to control disease. In some cases it seems preferable, for the benefit of bird life, actually to use insecticides and leave the habitat otherwise unchanged.

How important is pollution, by industry and pesticides, to wild life in Britain to-day? It is difficult to say, because we lack basic knowledge about so many species. We have accurate statistics about peregrines and a few other species of birds. We have fairly good records of some of the rarer plants. But when someone states that many of our previously plentiful butterflies have become scarce in recent years, it is difficult to be sure that this is a fact, and even more difficult to discover the reason. All these studies show how essential it is that we have a national stocktaking, and collect accurate data about as many species as is possible. We also need more research in the way climatic and other factors in the environment affect wild populations. Only when we have this information can we say with any certainty that man-made changes have had particular effects. We know that changes in farming practice, the removal of hedges, the ploughing of marginal land, must have important effects on wild life. Many valuable habitats have already been destroyed and more are disappearing. Some take a defeatist attitude, and say that everything is lost. The reverse is surely the case. There is still so much to save. But unchecked pollution is a real danger, an even greater danger than it would be if all the other pressures on the countryside did not also exist. Unless we control pollution, particularly the insidious effects of persistent poisonous substances, the losses may be irreversible. But pollution control by itself is valueless. Our need is for a more positive approach on the part of the whole nation, not just a few enthusiasts, often thought of as cranks, to the problem of wild life conservation in our crowded island.

BIBLIOGRAPHY

BIBLIOGRAPHY

It would be impossible to include references to all relevant books and papers. Those selected here have been chosen to illustrate particular fields. Authoritative summaries, which contain further references to detailed original work, have been included when possible, together with several reports of important conferences. Some original research publications, generally the first papers to illustrate particular important aspects, are also listed.

CHAPTER 1: INTRODUCTION

1.1 STAMP, L. D. (1955) *Man and the Land.* New Naturalist No. 31, Collins, London, pp. xvi+272.

1.2 STAMP, L. D. (1962) *The Land of Britain. Its Use and Misuse.* Third Edition, Longmans, Green, London, pp. viii+546.
These illustrate how Britain is changing because of the way the land is used. They set the scene for the study of the effect of different factors, including pollution.

1.3 YAPP, W. B. (1959) *The Effects of Pollution on Living Material.* Symposia of the Institute of Biology, No. 8, London, pp. xxii+1954.

1.4 GOODMAN, G. T., EDWARDS, R. W. and LAMBERT, J. M. (1965) *Ecology and The Industrial Society.* A Symposium of the British Ecological Society, Swansea, 13-16 April, 1964. Blackwell, Oxford, pp. 395.
Two important conferences which dealt with all types of environmental pollution.

1.5 THOMAS, W. J. (ED.) (1956) *Man's Role in Changing the Face of the Earth.* University of Chicago Press, pp. xxxviii+1193.
An international symposium summarising the global situation.

1.6 THE ENVIRONMENTAL POLLUTION PANEL (1965) *Restoring the Quality of our Environment.* Report of the President's Science Advisory Committee, November, 1965, pp. 317.
Summarises the situation in the U.S.A. in 1965, and includes 104 specific recommendations to reduce pollution.

1.7 MOORE, N. W. (1957) The Past and Present Status of the Buzzard in the British Isles. *British Birds* 50, 173-197.
Shows how the number of one species has fluctuated during the last 150 years.

When this book was in the press Garth Christian's book *Tomorrow's Countryside: The Road to the Seventies* (John Murray, London, pp. xii+229, 1966) appeared. It gives an excellent account of the way our countryside is changing, and suggests how our resources should be wisely used.

CHAPTER 2: AIR POLLUTION

There are few books dealing with this subject in a general way. 2.1, though on control, does give a good account of the main problems. Useful articles appear in 1.3. and 1.4. listed above, but it is noteworthy that 1.3. devotes 32 pages to air pollution, and 69 to fresh water, and 1.4. has 37 on air pollution, 91 on fresh water. This is a measure of the way biologists have neglected this subject.

Detailed records of atmospheric pollution and summaries of the literature appear in the "Atmospheric Pollution Bulletin," from the Warren Spring Laboratory, Ministry of Technology (formerly the Department of Scientific and Industrial Research).

2.1 GILPIN, A. (1963). *Control of Air Pollution.* Butterworth, London, pp. xvi+514

2.2 FORD, E. B. (1964) *Ecological Genetics.* Methuen, London, pp. xv+335.
Includes an account of industrial melanism and similar phenomena.

2.3 NATURAL HISTORY OF THE GARDEN OF BUCKINGHAM PALACE (1963). *Proc. S. Lond. ent. nat. Hist. Soc.* 1963, pp. 140.
This survey illustrates the richness of the flora and fauna which may be found even with the high atmospheric pollution of Central London.

2.4 FENTON, A. F. (1960) Lichens as indicators of atmospheric pollution. *Irish Naturalists Journal, 13,* pp. 153-158.

2.5 FENTON, A. F. (1964) Atmospheric pollution of Belfast and its relationship to the lichen flora. *Irish Naturalists Journal, 14,* 273-245.

CHAPTER 3: WATER POLLUTION

In contrast to air pollution, there is no lack of literature on water pollution.

3.1 MACAN, T. T. and WORTHINGTON, E. B. (1951) *Life in Lakes and Rivers.* The New Naturalist No. 15, Collins, London, xvi+272.
An introduction to the fresh-water habitat.

3.2 MACAN, T. T. (1963) *Freshwater Ecology.* Longmans, London, x+338.
A scientific account of the habitat.

3.3 HYNES, H. B. N. (1960) *The Biology of Polluted Waters.* Liverpool University Press, xiv+202.

3.4 JONES, J. R. ERICHSEN (1964) *Fish and River Pollution.* Butterworth, London, viii+203.
Detailed accounts of pollution.

3.5 PRAT, J. and GIRAND, A. (1964) *The Pollution of Water by Detergents.*
Organisation for Economic Co-operation and Development, pp. 86.
See also appropriate chapters in 1.3, 1.4, 1.5.

CHAPTER 4: RADIATION

4.1 HADDOW, A. (ED.) (1952) *Biological Hazards of Atomic Energy.* An Institute of Biology Symposium, Oxford University Press, pp. xi+235

4.2 AUGENSTEIN, L. G., MASON, R. and QUASTLER, H. (1964) *Advances in Radiation Biology,* Vol. 1. Academic Press, New York.

Various reports of the Medical Research Council deal with hazards to man. See also 1.3, 1.4.

CHAPTER 6: HERBICIDES

6.1 SALISBURY, E. (1961) *Weeds and Aliens.* The New Naturalist No. 43, Collins, London, pp. 384.
An authoritative account of the weed problem.

6.2 WOODFORD, E. K. and EVANS, S. A. (ED.) *Weed Control Handbook.* Blackwell, Oxford. Gives details of the properties and uses of most herbicides.

6.3 MINISTRY OF AGRICULTURE, FISHERIES AND FOOD. *Farm Sprayers and Their Use.* Bulletin No. 182, H.M.S.O., London, pp. 99.
Explains how herbicides should be applied.
Many of the references in Chapter 8 deal with herbicides as well as insecticides. See in particular 8.1 – 8.7.

6.4 MARTIN, H. (1964) *The Scientific Principles of Crop Protection.* Fifth Edition. Edward Arnold, London, viii+376.
Deals with all chemicals used to protect crops, including herbicides, fungicides and weedkillers.

6.5 MINISTRY OF AGRICULTURE, FISHERIES AND FOOD. *Agricultural Chemicals Approval Scheme. List of Approved Products 1965 for Farmers and Growers,* pp. 143.
Lists all approved herbicides, fungicides and insecticides.

6.6 YEMM, E. W. and WILLIS, A. J. (1962) *The effect of maleichydrazide* and 2,4, Dichlorophenoxyacetic acid on roadside vegetation. *Weed Research, 2,* 24-40.
A scientific assessment of a long-term experiment on herbicides used to control the growth of roadside verges.

CHAPTER 7: FUNGICIDES

Though widely used and potentially dangerous, there has been less study of the effects of fungicides than of insecticides in relation to wild life.
See 6.4, 6.5.

7.1 MARTIN, H. (ED.) *Insecticide and Fungicide Handbook.* Blackwell, Oxford.
Gives details of the properties and uses of most fungicides (and also of insecticides).

7.2 BORG, K., WANNTROP, H., ERNE, K. and HANKO, E. (1965) Kvicksilverförgiftningar bland vilt i Sverige. *Statens veterinärmedicinska amstalt,* Stockholm, pp 53+tables.
Mercury poisoning, possibly associated with fungicidal seed-dressings.

7.3 RAW, F. (1962) Studies of earthworm populations in orchards. *Ann. appl. Biol, 50,* 389-404.
Profound ecological effects of copper fungicides.

CHAPTER 8: INSECTICIDES AND INSECT CONTROL

There is now an immense literature of unequal quality. The fullest recent scientific assessment of the problem is given in 8.16. Only a selection of important sources is included. 6.2, 6.3, 6.4, 6.5 and 7.1 are also important and basic texts.

8.1   CARSON, R. (1963) *Silent Spring*. Hamish Hamilton, London, xxii+304.
      This book had the greatest public impact. It gives an advocate's view,
      one-sided, but, as regards insecticides, generally accurate. Some of the medi-
      cal details have not been generally corroborated.

8.2   RUDD, R. L. (1965) *Pesticides and the Living Landscape*. Faber and Faber, London,
      xiv+320.
      Another, but quite different, American account. This is an admirable sum-
      mary of the literature, considered critically from the ecological point of view.

8.3   COLEMAN-COOKE, J. (1965) *The Harvest that Kills*. Odhams, London, pp. 208.
      Has been described as "The British Silent Spring"; however, it does recog-
      nise the value to agriculture of insecticides, and weighs up the advantages
      and disadvantages of various means of pest control.

8.4   HUXLEY, E. (1965) *Brave New Victuals: An Inquiry into Modern Food Production*.
      Chatto and Windus, London, pp. 168.
      Includes a popular but accurate account of pesticide use in agriculture.

8.5   LAVERTON, S. (1962) *The Profitable Use of Farm Chemicals*. Oxford University
      Press, London, x+96.

8.6   CHEMICALS AND THE LAND IN RELATION TO THE WELFARE OF MAN. Report of
      Symposium held at Yorkshire (W.R.) Institute of Agriculture, Askham Bryan,
      York, on 12th, 13th, 14th April 1965, pp. 152.

8.7   FOOD SUPPLY AND NATURE CONSERVATION: A SYMPOSIUM. Cambridgeshire
      College of Arts and Technology, 1964.
      Critical accounts of chemical use, from various points of view.

8.8   JONES, F. G. W. and JONES, M. (1964) *Pests of Field Crops*. Edward Arnold, London,
      viii+406.

8.9   EDWARDS, C. A. and HEATH, G. W. (1964) *The Principles of Agricultural Entomology*.
      Chapman and Hall, London, xiv+418.

8.10 MASSEE, A. M. (1954) *The Pests of Fruits and Hops*, 3rd Ed. Crosby Lockwood,
      London, xvi+325.
      Useful textbooks, which include some accounts of spray use.

6.11 REVIEW OF THE PERSISTENT ORGANOCHLORINE PESTICIDES. Advisory Com-
      mittee on Poisonous Substances used in Agriculture and Food Storage. Minis-
      try of Agriculture, Fisheries and Food, 1964.

8.12 REVIEW OF THE PERSISTENT ORGANOCHLORINE PESTICIDES: SUPPLEMENTARY
      REPORT. The Advisory Committee on Pesticides and other Toxic Chemicals.
      Ministry of Agriculture, Fisheries and Food, pp. 8.

8.13 REPORT OF THE RESEARCH COMMITTEE ON TOXIC CHEMICALS. Agricultural
      Research Council, 1964, pp. 38.

8.14 SUPPLEMENTARY REPORT OF THE RESEARCH COMMITTEE ON TOXIC CHEMICALS.
      Agricultural Research Council, 1965.
      Official policy in Britain regarding pesticides. See also 6.2 and 7.1 in this
      connection.

8.15 EVALUATION OF THE TOXICITY OF PESTICIDE RESIDUES IN FOOD (1964). Food
      and Agricultural Organisation of the United Nations. World Health
      Organisation, pp. 172.
      A summary of toxicity, and of the amounts considered tolerable in human
      food.

8.16 MOORE, N. W. (ED.) (1966) Pesticides in the Environment and their Effects on Wild Life [Proceedings of a NATO Advanced Study Institute] Supplement to J. appl. Ecol., pp. 300.
This recent account summarises much of the work in Britain and elsewhere. It includes reference to all important work up to 1965.

8.17 CRAMP, S. and CONDER, P. (1965) *The fifth report of the Joint Committee of the British Trust for Ornithology and the Royal Society for the Protection of Birds on Toxic Chemicals, August 1963-July 1964.* Royal Society for the Protection of Birds, pp. 20.
Readers may also be interested in the previous four reports, which deal with bird deaths from 1956 onwards. These were among the first accounts of the effects of seed-dressings.

8.18 ANNUAL REPORTS, GAME RESEARCH ASSOCIATION
These Annual Reports contain original reports on the effects of pesticides on game birds.

The following are a selection of important research papers on work in Britain.

8.19 MURTON, R. K. and VIZOSO, M. (1963) Dressed cereal seed as a hazard to woodpigeons. *Ann. appl. Biol. 52*, 503-517.

8.20 RATCLIFFE, D. A. (1963) The status of the Peregrine in Great Britain. *Bird Study 10*, 56-90.

8.20 MOORE, N. W. and WALKER, C. H. (1964) Organic chlorine insecticide residues in wild birds. *Nature, Lond., 201*, 1072-1073.

8.22 LOCKIE, J. D. and RATCLIFFE, D. A. (1964) Insecticides and Scottish Golden Eagles. *British Birds, 57*, 89-102.

8.23 MOORE, N. W. and TATTON, J. O'G. (1965) Organochlorine insecticide residues in the eggs of sea birds. *Nature, Lond., 207*, 42-43.

8.24 MOORE, N. W. (1965) Pesticides and Birds – a review of the situation in Great Britain in 1965. *Bird Study, 12*, 222-252.

8.25 PRESTT, I. (1965) An inquiry into the recent breeding success of some of the smaller birds of prey and crows in Britain. *Bird Study, 12*, 196-221.

82.6 SLADEN, W. J. L., MENZIE, C. M. and REICHEL, W. L. (1966) DDT residues in Adelie Penguins and Crabeater Seals from Antarctica. *Nature, Lond., 210*, 670-673.
Evidence of the presence of DDT in the Antarctic.

CHAPTER 10: THE CONTROL OF VERTEBRATE PESTS

Much of the work on vertebrate pest control is done by commercial firms, and detailed accounts have not yet been published in some cases.

10.1 INFESTATION CONTROL. *Report of the Infestation Control Laboratory for 1962-64.* H.M.S.O. vi+100.
Report of the main laboratory for work on this subject.

10.2 THOMPSON, H. V. and WORDEN, A. N. (1956) *The Rabbit.* 'New Naturalist' Special Volume No. 13, Collins, London, xii+240.

10.3 MURTON, R. K. (1965) *The Wood-Pigeon.* 'New Naturalist' Special Volume No. 20, Collins, London, pp. 256.

10.4 SUMMERS-SMITH, D. (1964) *The House Sparrow*. 'New Naturalist' Special Volume No. 19, Collins, London, pp. xiv+270.
Accounts of three "pests," and their control.

CHAPTER 11: THE FUTURE - IMPROVEMENT OR DISASTER?

Most of the references given for Chapter 1 also deal with this subject. The sources listed here concern methods, the use of which might reduce pollution.

11.1 DE BACH, P. (ED.) (1964) *Biological Control of Insect Pests and Weeds*. Chapman and Hall, London, pp. xxiv+844.
The authoritative account; for the more advanced reader.

11.2 SWAN, L. A. (1964) *Beneficial Insects*. Harper and Row, New York and London, pp. xvii+429.
This book has a misleading title; it deals with all methods used in biological control. It is more popular in appeal than 1.1, but is equally accurate

11.3 STEINHAUS, E. A. (1963) *Insect Pathology*. Academic Press, London. Vol. 1, xvii+661; Vol. 2, xiv+689.

11.4 PAINTER, R. H. (1951) *Insect Resistance in Crop Plants*. Macmillan, New York, pp. x+520.

11.5 BERTRAM, D. S. (1964) Entomological and Parasitological Aspects of Vector Chemo-sterilization. *Trans. Roy. Soc. Trop. Med. Hyg. 58*, 296-317.

11.6 BARNES, J. M. (1964) Toxic hazards and the use of insect chemosterilants. *Trans. Roy. Soc. Trop. Med. Hyg., 58*, 327-332.

11.7 FAHMY, O. G. and FAHMY, M. J. (1964) The Chemistry and Genetics of the Alkylating Chemosterilants. *Trans. Roy. Soc. Trop. Med. Hyg., 58*, 318-326.

11.8 HILLS, LAWRENCE D. (1964) *Pest Control without Poisons*. Henry Doubleday Research Association, pp. 64.
An unorthodox approach to pest control.

APPENDIX

Composition of some chemicals described in the text

Chapter 6 Herbicides

DNOC	2–methyl–4, 6–dinitrophenol
Dinoseb	2, 4–dinitro–6–s–butylphenol
MCPA	4–chloro–2–methylphenoxyacetic acid
2, 4–D	2, 4–dichlorophenoxyacetic acid
2, 4, 5–T	2, 4, 5-trichlorophenoxyacetic acid
Maleic hydrazide	1, 2–dihydropyridazine–3, 6–dione
Simazine	2–chloro–4, 6–bisethylamino–1, 3, 5–triazine
Monuron	N'–(4–chlorophenyl)–NN–dimethylurea
Dalapon	2, 2–dichloropropionic acid
Paraquat (–dichloride)	1, 1'–dimethyl–4, 4'–bipyridylium dichloride

Chapter 8 Insecticides

DNOC	2–methyl–4, 6–dinitrophenol
Parathion	OO–diethyl O–p–nitrophenyl phosphorothioate
TEPP	Bisdiethylphosphoric anhydride
Phorate	OO–diethyl S–(ethylthiomethyl) phosphorodithioate
Disulfoton	OO–diethyl S–2–(ethylthio)ethyl phosphorodithioate
Phosphamidon	Dimethyl 2–chloro–2–diethylcarbamoyl–1–methylvinyl phosphate
Dichlorvos	2, 2–dichlorovinyl dimethyl phosphate
Demeton-methyl	OO–dimethyl O–2–(ethylthio)ethyl phosphorothioate and S–isomer
Menazon	OO–dimethyl S–(4, 6–diamino–1, 3, 5–triazin–2–yl methyl) phosphorodithioate
Malathion	OO–dimethyl S–1, 2–di–(ethoxycarbonyl)ethyl phosphorodithioate
Schradan	Bis–NNN'N'–tetramethyl phosphorodiamidic anhydride

DDT   2, 2-bis(p-chlorophenyl)-1, 1, 1-trichloroethane

DDD (syn. TDE,
  Rhothane)   2, 2-bis(p-chlorophenyl)-1, 1-dichloroethane

Benzene
hexachloride
(syn. Lindane,
  gamma BHC) γ-1, 2, 3, 4, 5, 6-hexachloro*cyclo*hexane

Aldrin   1, 2, 3, 4, 10, 10-hexachloro-1, 4, 4a, 5, 8, 8a-
hexahydro-*exo*-1, 4-*endo*-5, 8-
dimethanonaphthalene

Dieldrin   1, 2, 3, 4, 10, 10-hexachloro-6, 7-epoxy-1, 4,
4a, 5, 6, 7, 8-8a, octahydro-*exo*-1, 4-*endo*-5,
8-dimethanonaphthalene

Heptachlor   1, 4, 5, 6, 7, 10, 10-heptachloro-4, 7, 8, 9-
tetrahydro-4, 7-*endo*-methyleneindene

Endosulfan   6, 7, 8, 9, 10, 10-hexachloro-1, 5, 5a, 6, 9, 9a-
hexahydro-6, 9-methano-2, 4, 3-benzo-
dixoathiepin-3-oxide

Endrin   1, 2, 3, 4, 10, 10-hexachloro-6, 7-epoxy-1, 4,
4a, 5, 6, 7, 8, 8a-octahydro-*exo*-1, 4-*exo*-5,
8-dimethanonaphthalene

INDEX

215